Biohacking

Biohacking

Carter Gray
Sarah Fleming

Kruger Brentt
Publishers
2026

Kruger Brentt Publishers UK. LTD.
Company Number 9728962

Regd. Office: 68 St Margarets Road, Edgware, Middlesex HA8 9UU

ISBN: 978-1-78715-660-9

For information on all our publications visit our website at http://krugerbrentt.com/

Preface

Biohacking has emerged as one of the most dynamic intersections of science, technology, and personal wellness, inviting individuals to actively participate in shaping their own biological potential. As new discoveries in genetics, neuroscience, and metabolic sciences move rapidly from research labs to everyday practice, the field offers both exciting opportunities and critical responsibilities. This book introduces readers to the foundations of biohacking with clarity and purpose, presenting its core ideas, evolving methodologies, and transformations in human–technology interaction.

Bringing together developments from DIY biology spaces, quantified-self communities, and contemporary biomedical innovation, the text highlights how thoughtfully applied biological insight can enhance performance, resilience, and well-being. At the same time, it underscores the need for scientific literacy, ethical reflection, and a respect for safety as experimentation becomes increasingly accessible. Designed to guide readers through the possibilities and limitations of this expanding landscape, the book aims to build informed, responsible, and confident practitioners in the science of self-optimization.

Carter Gray

Sarah Fleming

Contents

CHAPTER 1

Introduction to Biohacking

1.1 WHAT IS BIOHACKING?

Biohacking is a contemporary movement that combines biology, technology, and self-experimentation to optimize human health, performance, and longevity. It is often described as the "do-it-yourself biology" movement, where individuals use a variety of techniques ranging from nutritional adjustments to advanced technological interventions in order to enhance their physical and mental capacities. At its core, biohacking is based on the belief that the human body can be modified, upgraded, and transformed beyond its natural limits through intentional interventions. The practice has gained momentum in recent years due to the rapid progress in biotechnology, neuroscience, and digital health tools. While some biohacking techniques are relatively simple, such as dietary changes or mindfulness training, others are highly sophisticated, involving genetic modification, nootropics, or implanted devices.

1.1.1 Historical Background of Biohacking

The roots of biohacking can be traced to traditional practices of human optimization. Ancient civilizations practiced forms of biohacking through herbal medicine, fasting, meditation, and physical training, all designed to strengthen the body and mind. However, the modern term "biohacking" began to emerge in the early 2000s when scientific knowledge and accessible technology converged. It became associated with the broader maker and hacker movements, where enthusiasts sought to take science out of laboratories and apply it to their daily lives. Early biohackers experimented with nutrition, sleep cycles, and exercise regimens, while others began tinkering with more advanced biological tools such as genetic editing technologies. Over time, biohacking evolved into a diverse field with distinct subcultures, from those focused on lifestyle optimization to those experimenting with body implants and synthetic biology.

1.1.2 Kinds of Biohacking

Biohacking can be broadly categorized into several approaches. One of the most common is nutritional biohacking, which involves modifying diet, supplements, and fasting patterns to improve health and longevity. Another form is nootropic biohacking, where individuals use cognitive enhancers, such as natural herbs or synthetic compounds, to improve memory, focus, and creativity. Genetic biohacking is another advanced category, where individuals attempt to alter their genes using technologies like CRISPR, although this area remains controversial and highly regulated.

A more visible form of biohacking is the practice of "grinders," who implant chips or devices into their bodies to gain new abilities, such as unlocking doors with implanted RFID tags or monitoring glucose levels with embedded sensors. Finally, there is psychological and behavioral biohacking, which involves techniques like meditation, neurofeedback, and sleep optimization to enhance mental well-being and performance. These diverse approaches show that biohacking is not a single discipline but a spectrum of practices united by the desire for self-improvement through biological manipulation.

1.1.3 Science behind Biohacking

Biohacking is rooted in scientific disciplines such as biology, physiology, neuroscience, and nutrition. The human body functions through complex systems involving cellular processes, metabolic pathways, and neurological signaling. By intervening in these processes, biohackers attempt to optimize outcomes that range from faster muscle growth to sharper mental clarity. For instance, fasting triggers a process called autophagy, where cells recycle damaged components, potentially slowing aging.

Similarly, nootropics work by influencing neurotransmitters like dopamine and serotonin, which regulate mood, motivation, and cognitive performance. Genetic biohacking, though still experimental, is based on the understanding of DNA sequences and gene expression, where small changes in genetic codes could have profound impacts on health. Wearable technology, another pillar of biohacking, integrates data science and physiology, enabling real-time monitoring of heart rate, sleep cycles, and blood oxygen levels. This data-driven approach allows individuals to experiment with their bodies in measurable ways, embodying the scientific spirit of trial and observation.

1.1.4 Popular Biohacking Practices

In the mainstream, some biohacking practices have become widely accepted. Intermittent fasting, for instance, is embraced by millions as a method for weight management and metabolic health. Ketogenic diets, which focus on high-fat and low-carbohydrate intake, are another form of dietary biohacking that influences energy metabolism. Cold exposure and ice baths have also become popular as biohacks that stimulate circulation, reduce inflammation, and enhance recovery. Meditation and mindfulness techniques represent psychological biohacking strategies that are linked with reduced stress, improved focus, and better emotional regulation. In addition, the growing market of supplements and nootropics provides individuals with various options for boosting energy, concentration, or sleep quality. The rise of wearable devices such as smartwatches and fitness trackers has further fueled biohacking by providing individuals with immediate feedback on how their interventions impact their physiology. These practices highlight how biohacking has moved from fringe experimentation to mainstream health and wellness culture.

1.1.5 Risks and Ethical Concerns of Biohacking

Despite its appeal, biohacking carries significant risks and raises ethical questions. Some practices, particularly genetic modification and unregulated chemical experimentation, may result in unintended health consequences. For example, self-administered genetic engineering without professional oversight can cause harmful mutations or immune responses. The use of untested nootropics may also lead to dependency, side effects, or unknown long-term impacts on brain health.

The biohacker movement has also been criticized for its lack of standardization and scientific rigor, as many self-experiments are anecdotal rather than systematically validated. On an ethical level, there are concerns about biohacking's implications for human identity and inequality. If advanced biohacking technologies become accessible only to wealthy individuals, they may create new forms of social disparity by giving some people enhanced abilities over others. Additionally, the idea of merging machines with human biology through implants and augmentation raises questions about what it means to be human. These risks highlight the importance of responsible experimentation and regulatory oversight in the biohacking community.

1.1.6 Biohacking and Longevity

One of the most ambitious goals of biohacking is the extension of human lifespan and healthspan. Proponents argue that aging is a biological process that can be slowed, and in some cases reversed, through targeted interventions. Techniques such as caloric restriction, fasting, and supplementation with compounds like resveratrol or NAD+ boosters are believed to support cellular health and reduce the effects of aging.

Cutting-edge biohacking also explores the potential of gene therapies to switch off aging-related genes or to regenerate tissues. Wearable devices and continuous health monitoring allow individuals to detect early signs of disease, making preventive health strategies more effective. While the dream of radical life extension remains speculative, biohacking has already influenced the growing field of longevity science, where researchers and enthusiasts alike are searching for methods to extend human vitality well beyond current limits.

1.2 HISTORY AND EVOLUTION OF THE BIOHACKING MOVEMENT

1.2.1. Precursors and Philosophical Foundations

The intellectual roots of biohacking stretch back to the transhumanist ideas that began to crystallize in the mid-20th century. Thinkers like Julian Huxley, who coined the term "transhumanism," and FM-2030 (formerly F.M. Esfandiary) proposed that humans could and should use technology to transcend biological limitations. This philosophy provided the ultimate "why" for biohacking. Simultaneously, the DIY (Do-It-Yourself) ethic, prevalent in computing and hardware cultures, provided the "how." The garage inventors and computer hackers of the 1970s and 80s demonstrated that complex systems could be understood, modified, and improved by passionate amateurs working outside corporate or academic structures. The convergence of transhumanism's ambitious goals with the DIY movement's pragmatic, accessible methods created the fertile ground from which biohacking would grow.

1.2.2.Quantified Self: The First Wave

The earliest and most widespread manifestation of the biohacking movement was the "Quantified Self" (QS), which emerged in the late 2000s. Pioneered by technology writers Kevin Kelly and Gary Wolf, QS was focused on self-knowledge through self-tracking. This was not about abstract curiosity but about using data to make precise, personalized interventions. The movement was enabled by the proliferation of consumer wearable technology like the Fitbit, Apple Watch, and a myriad of smartphone apps that allowed individuals to easily monitor metrics such as steps taken, heart rate, sleep cycles, and calorie intake. Biohackers began experimenting on themselves, using this data to optimize their daily routines. They would adjust their diet based on blood glucose monitors, modify their sleep environment based on sleep stage data, and tailor workout regimens based on heart rate variability. This data-driven approach established a foundational principle of biohacking: the body is a system that can be measured, analyzed, and optimized for enhanced performance and well-being.

1.2.3. Grinders and DIY Biology: The Hardware Frontier

While the Quantified Self focused on data collection, a more radical and hardware-oriented branch of biohacking was developing in parallel: the "grinder" movement. Grinders take a more literal and invasive approach to hacking the human body, often involving body modification and the implantation of devices. Early examples included implanting small magnets in fingertips to sense electromagnetic fields or embedding RFID chips to unlock doors and store data. This subculture pushed the boundaries of what was considered possible and acceptable, operating in a legal and ethical grey area. Around the same time, the DIYBio (Do-It-Yourself Biology) movement began to form, led by individuals like Mackenzie Cowell and Jason Bobe. They established community laboratories, or "hackerspaces for biology," such as Genspace in New York City.

These labs provided citizen scientists with access to equipment like PCR machines and centrifuges, enabling them to conduct experiments in genetics, microbiology, and synthetic biology outside of traditional academia and industry. This strand of biohacking emphasized open-source protocols, public education, and the radical democratization of biological tools.

1.2.4. Mainstream Adoption and Commercialization

Throughout the 2010s, biohacking began to shed its fringe image and entered the mainstream wellness and performance lexicon. Influential figures and authors, such as Dave Asprey with his Bulletproof Coffee and biohacking conferences, played a significant role in popularizing the term and its practices. The language of optimization resonated deeply with high-performing professionals in Silicon Valley and the broader tech industry, where a premium was placed on cognitive function and productivity.

This led to an explosion of commercial products and services marketed to aspiring biohackers, including nootropic supplements, red light therapy devices, cryotherapy chambers, and advanced genetic testing kits. While this commercialization made biohacking more accessible, it also created a divide between evidence-based practices and speculative, sometimes pseudoscientific, claims. The movement became a multi-billion dollar industry, blending cutting-edge science with wellness culture.

1.2.5. Cutting Edge: CRISPR and the Genetic Frontier

The most recent and profound evolution within biohacking is its foray into genetic engineering. The development of CRISPR-Cas9, a cheap, precise, and accessible gene-editing tool, has placed power that was once the sole domain of well-funded research institutions into the hands of amateurs. So-called "biohackers" or "garage biologists" now experiment with CRISPR kits, often purchasing them online to perform basic genetic engineering on bacteria or yeast.

This development is the ultimate expression of the movement's democratizing ethos but also raises the most serious ethical and safety concerns. The infamous case of He Jiankui, who created the first gene-edited babies, though not a biohacker in the traditional sense, exemplified the terrifying potential for misuse. The community now grapples with a critical question: how to foster open innovation and learning while establishing responsible safeguards to prevent reckless or malicious experiments that could have irreversible consequences for individuals and public health.

1.3 CATEGORIES OF BIOHACKERS

1.3.1 DIY Biohackers

The DIY biohacking movement represents a fascinating intersection of science, technology, and personal autonomy. It challenges traditional notions of medical authority, research ethics, and human potential. While the movement has the potential to democratize science and lead to groundbreaking discoveries, it also carries significant risks that cannot be ignored.

As biohacking continues to grow, it will be crucial for policymakers, scientists, and biohackers themselves to engage in dialogue about safety standards, ethical boundaries, and regulatory frameworks. The goal should be to foster innovation while minimizing harm, ensuring that the pursuit of human enhancement does not come at the cost of public health or societal well-being. Ultimately, the rise of DIY biohacking forces us to reconsider what it means to be human in an age where biology is no longer a fixed condition but a malleable frontier of personal experimentation.

The future of DIY biohacking is likely to be shaped by advancements in biotechnology, artificial intelligence, and the growing accessibility of scientific tools. As gene-editing technologies become more affordable and user-friendly, more individuals may attempt to modify their own DNA, potentially leading to breakthroughs in personalized medicine but also increasing risks of unintended genetic consequences.

The integration of biohacking with other emerging technologies, such as brain-computer interfaces or nanotechnology, could further blur the line between human and machine, raising profound philosophical questions about identity and humanity. On the positive side, the biohacking movement could accelerate medical research by fostering a culture of open-source science, where citizen researchers contribute to discoveries outside traditional academic and corporate structures. However, without proper safeguards, the movement could also lead to dangerous experimentation, widening health disparities, and even the emergence of a new form of biological inequality where only the privileged can afford enhancements.

Common Biohacking Practices and Technologies

DIY biohackers employ a wide range of techniques, from relatively safe and well-documented methods to highly experimental and risky procedures. One of the most accessible forms of biohacking involves nutritional and lifestyle interventions, such as ketogenic diets, cold exposure therapy, and sleep optimization using wearable devices. More advanced biohackers experiment with nootropic drugs, peptides, and even self-administered hormone therapies to enhance cognitive function and physical performance. A particularly bold subset of biohackers, known as grinders, engage in body modifications by implanting RFID chips, magnets, or even LED lights under their skin to interact with technology in novel ways. Some have gone further, attempting to enhance their senses or even achieve night vision through experimental procedures. The most controversial area of biohacking involves genetic modification, where individuals use CRISPR kits to edit their own DNA, raising significant ethical and safety concerns.

Ethical and Safety Concerns in DIY Biohacking

While the DIY biohacking movement empowers individuals to take control of their own biology, it also operates in a legal and ethical gray area. Unlike pharmaceutical drugs or medical devices, which undergo rigorous testing and regulatory approval, many biohacking experiments lack scientific validation, leading to potential health risks. Self-experimentation with unproven gene therapies, unregulated implants, or experimental drugs can result in unintended consequences, including infections, immune reactions, or long-term genetic damage. Additionally, the lack of oversight means that failed experiments may go unreported, making it difficult to assess the true risks of certain biohacking practices. Ethically, the movement raises questions about informed consent, particularly when biohackers share their experiments online, potentially encouraging others to follow unsafe practices. There is also the concern of biohacking being used for non-therapeutic enhancements, such as cognitive or physical augmentation, which could exacerbate social inequalities if only accessible to the wealthy.

Legal Landscape and Regulatory Challenges

The legal status of DIY biohacking varies significantly across different jurisdictions. In some countries, certain biohacking practices, such as gene editing or implanting unapproved devices, may be explicitly prohibited under medical or biotechnology regulations. However, enforcement is often difficult, as many biohackers operate in underground or online communities where activities are hard to monitor. In the United States, the FDA regulates medical devices and gene therapies, but personal use of CRISPR kits or self-implantation of non-medical devices exists in a regulatory blind spot. Some biohackers argue that they should have the right to modify their own bodies, comparing it to other forms of personal freedom, such as tattooing or body piercing. However, regulators and medical professionals warn that unchecked biohacking could lead to public health risks, particularly if modified genes or experimental treatments spread beyond the individual. The legal ambiguity surrounding biohacking will likely continue to evolve as technology advances and more people engage in self-experimentation.

1.3.2 Grinder Biohackers

Grinder biohackers represent one of the most daring and unconventional branches of the biohacking movement. Driven by the desire to transcend biological limitations and embrace technological evolution, grinders experiment with body implants, sensory enhancements, and neural interfaces.

Their philosophy emphasizes personal autonomy, self-experimentation, and the rejection of natural constraints. While their practices carry serious risks and face ethical scrutiny, they also embody the spirit of innovation and challenge conventional notions of what it means to be human. As science and technology continue to advance, the grinder movement provides a glimpse into a future where the boundaries between biology and technology are increasingly blurred. Whether viewed as reckless or visionary, grinders stand as a provocative reminder of humanity's relentless pursuit of self-transformation and progress.

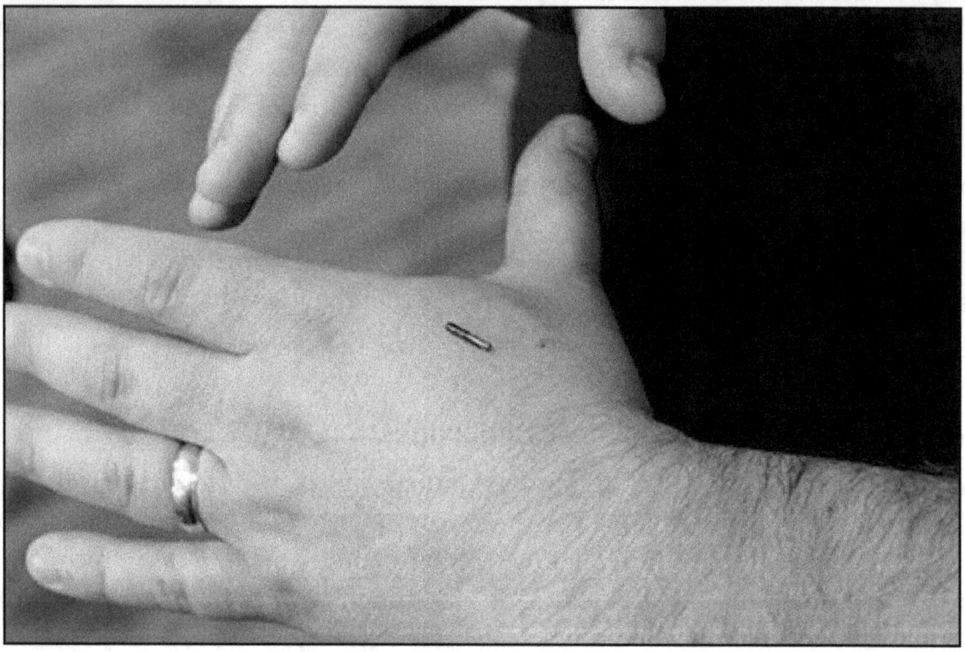

Grinder Biohackers

The future of grinder biohacking will likely be shaped by advances in biotechnology, nanotechnology, and neural engineering. As implantable devices become smaller, safer, and more powerful, the line between grinder experimentation and mainstream medical technology may blur.

Already, medical implants such as pacemakers, cochlear implants, and continuous glucose monitors reflect aspects of the grinder vision, though developed under strict regulation. In the future, it is possible that some grinder innovations, such as subdermal chips or sensory augmentation, may be adopted more widely as society becomes comfortable with body-integrated technology. However, the grinder community will continue to face challenges from regulatory authorities, ethical debates, and public skepticism. Whether grinders remain a fringe subculture or evolve into a mainstream movement, their contributions to the dialogue on human enhancement are undeniable.

Origins of the Grinder Movement

The grinder movement emerged in the early 2000s, drawing inspiration from both the biohacking community and the transhumanist philosophy that envisions the enhancement of human capabilities through technology. Early grinders were often influenced by cyberpunk literature and science fiction, which portrayed futures where humans and machines seamlessly integrated. Some of the first documented grinders experimented with radio-frequency identification (RFID) chips implanted in their hands, allowing them to unlock doors or start vehicles without keys. These initial experiments highlighted both the creative ingenuity and the risks of do-it-yourself body modification. Over time, online communities began to form, where grinders shared techniques, tools, and experiences, building a decentralized movement outside of traditional scientific institutions.

Philosophy of Grinders

At the heart of grinder biohacking lies the philosophy that the human body is a platform for innovation. Grinders reject the notion that humans must passively accept the limits set by biology and instead argue for active control over evolution. This belief is closely tied to transhumanism, which advocates the use of technology to enhance physical, cognitive, and sensory abilities. For grinders, body modification is not merely about aesthetics but about functionality, utility, and personal empowerment. They envision a future where implanted devices can improve memory, expand sensory perception, or provide constant health monitoring. Unlike mainstream medicine, which often prioritizes safety and regulation, grinders emphasize personal freedom and experimentation, even when it involves significant risks.

Methods and Techniques

Grinders use a wide range of methods to augment their bodies. One of the most common practices is the implantation of RFID or near-field communication (NFC) chips, which can be programmed to perform tasks such as unlocking smartphones, storing personal data, or replacing credit cards. Another popular modification involves the implantation of magnets in the fingertips, which allows individuals to sense electromagnetic fields and interact with metallic objects in unique ways. Some grinders experiment with light-emitting implants, such as subdermal LEDs, which can create glowing tattoos beneath the skin. More ambitious projects include neural implants designed to interface with computers, exoskeletal attachments to enhance strength, and biochemical modifications to alter bodily functions. These techniques often involve improvised surgical procedures conducted outside formal medical settings, raising questions about safety, legality, and ethics.

Role of DIY Science in Grinding

Grinder biohacking thrives on the do-it-yourself ethos. Because mainstream medical institutions and regulatory bodies often do not permit such radical experimentation, grinders operate largely outside traditional frameworks. Many grinders teach themselves surgical techniques, sterilization methods, and basic biomedical engineering. Online forums, workshops, and community laboratories provide platforms for sharing knowledge and resources. This DIY approach is both empowering and dangerous, as it enables innovation without institutional barriers but also exposes practitioners to significant health risks. By conducting experiments on themselves, grinders embody the philosophy of radical self-experimentation, positioning their bodies as both laboratory and test subject.

Benefits and Aspirations of Grinder Biohackers

Supporters of the grinder movement argue that their experiments pave the way for revolutionary innovations in human enhancement. The implantation of chips and sensors can create new forms of interaction with the digital world, offering convenience and expanded abilities. For example, implanted magnets allow individuals to perceive invisible electromagnetic forces, effectively giving them a "sixth sense."

Grinders aspire to develop technologies that could improve memory, regulate emotions, or monitor health in real time. Some dream of neural interfaces that would allow direct communication between the brain and computers, potentially transforming learning, communication, and creativity. These aspirations align with a vision of the future where technology is seamlessly integrated into human biology, creating a new stage in human evolution.

Risks and Challenges of Grinder Biohacking

The grinder movement faces significant risks and challenges. One of the most pressing concerns is safety. Many grinder procedures are conducted in non-sterile environments without medical supervision, increasing the risk of infection, rejection, or long-term complications. The use of untested devices can also lead to unforeseen health consequences, such as tissue damage, nerve injury, or chronic pain. Legal and ethical issues further complicate the movement, as most countries have strict regulations regarding surgical procedures and medical devices. Critics argue that grinder biohacking is reckless and dangerous, while proponents counter that innovation often requires risk-taking. Additionally, the lack of scientific validation for many grinder experiments makes it difficult to assess their effectiveness or long-term impact.

Grinders and the Transhumanist Vision

Grinders are often seen as pioneers of the transhumanist vision, where technology and biology merge to create a post-human future. Transhumanism advocates for using science and technology to overcome biological limitations, such as aging, disease, and cognitive boundaries. Grinders embody this philosophy through direct, hands-on experimentation with their own bodies. They serve as early adopters and testers of technologies that may one day become mainstream. While current grinder modifications may appear crude compared to the advanced visions of transhumanists, they represent the first steps toward integrating technology into human biology. In this sense, grinders act as the vanguard of a movement that could fundamentally reshape the future of humanity.

Social and Cultural Impact of Grinders

The grinder movement has attracted both fascination and skepticism in popular culture. Media often portrays grinders as fringe experimenters who push the boundaries of science fiction into reality. Their practices raise important questions about identity, privacy, and the definition of humanity in an age of increasing technological integration. The visibility of grinders has also influenced debates about bioethics, transhumanism, and the limits of human enhancement. Some critics fear that widespread adoption of grinder technologies could create new forms of inequality, where only those with access to implants gain enhanced abilities. Others view grinders as visionaries who challenge society to reconsider outdated notions of human potential. Regardless of perspective, grinders have succeeded in sparking dialogue about the role of technology in shaping the human future.

1.3.3 Nutritional and Lifestyle Hackers

Nutritional and lifestyle hacking represents a fascinating convergence of science, technology, and personal experimentation, offering individuals unprecedented control over their health and performance. While the movement has the potential to revolutionize how we approach well-being, it also carries risks that must be carefully managed. The key to successful and sustainable lifestyle hacking lies in a balanced approach—one that leverages data and science while remaining mindful of individual differences and the broader context of health.

As the field continues to evolve, it will be essential for practitioners to stay informed, critically evaluate new trends, and prioritize long-term well-being over short-term gains. Ultimately, the goal of nutritional and lifestyle hacking should not be perfection but rather a deeper understanding of one's own biology and the ability to make informed choices that enhance both performance and quality of life. As technology continues to advance, the field of nutritional and lifestyle hacking is likely to become even more sophisticated. The rise of personalized nutrition, driven by AI and genetic testing, will allow for even more tailored dietary and supplement recommendations based on an individual's unique biology. The integration of real-time biosensors and closed-loop systems, where devices automatically adjust environmental or dietary inputs based on biological feedback, could make optimization seamless and highly individualized. There is also growing interest in the intersection of nutritional hacking with longevity science, where researchers and biohackers alike are exploring how specific dietary patterns and supplements may influence aging and age-related diseases. However, as the field evolves, it will be crucial to balance innovation with responsible experimentation, ensuring that the pursuit of optimization does not come at the cost of long-term health or psychological well-being. The future may see a greater emphasis on holistic approaches, where mental health, social connections, and sustainability are given as much importance as physical and cognitive performance.

Science behind Nutritional Hacking

Nutritional hacking is built on the principle that food is not just fuel but a powerful tool for optimizing biological function. Many lifestyle hackers adopt diets that go beyond basic macronutrient balance, instead focusing on how specific foods, nutrients, and eating patterns influence metabolism, brain function, and cellular health. One of the most popular nutritional hacking strategies is intermittent fasting, which involves cycling between periods of eating and fasting to enhance metabolic flexibility, improve insulin sensitivity, and promote autophagy—the body's natural process of cellular repair. Another widely adopted approach is the ketogenic diet, which shifts the body's primary energy source from carbohydrates to fats, leading to improved mental clarity and sustained energy levels for some individuals. Beyond macronutrient manipulation, nutritional hackers often experiment with micronutrients, adaptogens, and nootropic supplements to fine-tune cognitive performance and stress resilience. The use of continuous glucose monitors (CGMs) has also become a trend among biohackers, allowing them to track how different foods affect their blood sugar levels in real time, enabling highly personalized dietary adjustments.

Lifestyle Optimization: Beyond Diet and Exercise

While nutrition is a cornerstone of lifestyle hacking, true optimization extends into other areas of daily life, including sleep, stress management, and environmental factors. Sleep hacking, for instance, has become a major focus for those seeking to maximize recovery and cognitive function. Techniques such as polyphasic sleep, blue light reduction, and temperature regulation are employed to improve sleep quality and efficiency. Some lifestyle hackers use wearable devices like the Oura Ring or Whoop strap to monitor sleep stages, heart rate variability, and recovery metrics, allowing them to adjust their routines based on real-time data. Stress management is another critical component, with practices like cold exposure, meditation, and breathwork being used to modulate the nervous system and enhance resilience. Environmental hacking also plays a role, with individuals optimizing their living and working spaces for factors like air quality, lighting, and electromagnetic field (EMF) exposure to minimize biological stressors. The goal is to create an environment that supports optimal physiological and psychological function, reducing unnecessary strain on the body and mind.

Role of Technology in Lifestyle Hacking

Technology has been a game-changer in the world of nutritional and lifestyle hacking, providing tools that allow individuals to quantify and optimize their biological processes with unprecedented precision. Wearable fitness trackers, smart scales, and metabolic analyzers provide continuous feedback on key health metrics, enabling users to make data-driven adjustments to their routines.

Apps like Cronometer and MyFitnessPal allow for detailed macronutrient and micronutrient tracking, while platforms like Zero and Life Fasting Tracker help users implement and monitor fasting protocols. Advanced blood testing services, such as those offered by InsideTracker and Nutrisense, provide deep insights into biomarkers like cholesterol, inflammation levels, and vitamin deficiencies, allowing for highly personalized nutritional interventions. Even genetic testing services like 23andMe and Nutrigenomix are being used to tailor diets and supplement regimens based on individual genetic predispositions. The integration of artificial intelligence and machine learning into these tools is further refining the ability to predict how specific lifestyle changes will impact an individual's health, making optimization more precise and accessible.

Controversies and Ethical Considerations

Despite the potential benefits of nutritional and lifestyle hacking, the movement is not without controversy. Critics argue that the relentless pursuit of optimization can lead to obsessive behaviors, where individuals become overly fixated on metrics and lose sight of overall well-being.

There is also concern about the long-term effects of extreme dietary practices, such as prolonged fasting or very low-carb diets, which may not be suitable for everyone and could have unintended health consequences. The use of unregulated supplements and nootropics raises additional ethical and safety questions, as many of these substances lack rigorous clinical testing. Furthermore, the accessibility of advanced biohacking tools and technologies can create disparities, where only those with financial resources can afford the latest optimizations, potentially widening health inequalities. There is also the risk of misinformation, as anecdotal success stories in online communities may not translate to universal benefits, leading some individuals to adopt practices that are not scientifically validated or appropriate for their unique physiology.

1.4 BIOHACKING AND THE QUANTIFIED SELF MOVEMENT

1.4.1 Convergence of Self and System: Biohacking and the Quantified Self

The human body, long treated as a mysterious black box governed by intuition and generalized medical advice, is now being approached as a system to be decoded, analyzed, and optimized. This fundamental shift in perspective is driven by the powerful synergy between two interconnected movements: biohacking and the Quantified Self. While biohacking provides the overarching philosophy of taking control of one's own biology, the Quantified Self movement supplies the foundational methodology—a rigorous, data-driven approach to self-knowledge. Together, they represent a paradigm where life is an experiment and the individual is both the subject and the scientist, leveraging technology to pursue enhanced performance, health, and well-being.

1.4.2 Philosophical Bedrock: From Hacker Ethos to Self-Knowledge

Biohacking is born from the marriage of the transhumanist vision—the belief in using technology to transcend human limitations—and the hands-on, iterative ethos of the hacker culture. It applies the principles of open-source information, DIY experimentation, and rapid prototyping to the human organism.

The Quantified Self, or QS, is a central pillar of this philosophy. Coined by Wired magazine editors Gary Wolf and Kevin Kelly in 2007, the Quantified Self movement is predicated on a simple but radical question: "What can you learn about yourself through numbers?" Its mantra, "self-knowledge through self-tracking," moves beyond vague feelings and anecdotal evidence, insisting that objective data is the key to unlocking a deeper understanding of one's physical and cognitive functions. This is not passive observation but active interrogation, where every aspect of daily life becomes a potential data point in a grand project of self-optimization.

1.4.3 Technological Catalyst: Wearables and the Data Stream

The rise of the Quantified Self would have remained a niche concept for a small group of tech enthusiasts without the parallel explosion of consumer biotechnology. The development of sophisticated, affordable, and user-friendly wearable sensors provided the essential tools to turn philosophy into practice. The launch of devices like the Fitbit in 2009 and the subsequent proliferation of smartwatches, most notably the Apple Watch, created a seamless pipeline of biometric data. These devices automated the tracking of once-elusive metrics such as step count, heart rate, sleep duration and quality, and blood oxygen levels. This ecosystem expanded to include smartphone apps for logging food intake and mood, connected scales that measure body composition, and even continuous glucose monitors (CGMs) repurposed for use by non-diabetics seeking to understand their metabolic responses to food. This democratization of monitoring technology removed the barrier of tedious manual entry, providing a constant, passive stream of data that forms the empirical foundation for biohacking interventions.

1.4.4 Practice of Optimization: From Data to Intervention

The true essence of the biohacking-Quantified Self synergy lies in the closed feedback loop of measurement, analysis, and intervention. A biohacker does not collect data for its own sake; they use it to inform precise, personalized changes to their lifestyle. For instance, an individual tracking their sleep with a wearable might notice that their deep sleep stages are consistently low.

They might then form a hypothesis: that exposure to blue light from screens in the evening is disrupting their circadian rhythm. Their intervention would be to implement a strict digital curfew, using blue-light-blocking glasses after sunset. They would then monitor their sleep data over the following weeks to see if the metric improves, thus validating or invalidating their hypothesis. This scientific method applied to the self can extend to nutrition, using a CGM to identify foods that cause undesirable blood sugar spikes; to cognitive performance, using brainwave-sensing headbands to optimize meditation or focus; and to physical fitness, using heart rate variability (HRV) data to tailor workout intensity and recovery periods. Each experiment is designed to move a specific metric in a desired direction.

1.4.5 Mainstream Ascent: From Niche Community to Global Phenomenon

What began in small meet-up groups where early adopters shared their self-tracking spreadsheets has now permeated mainstream culture. The language of the Quantified Self has been adopted by corporate wellness programs, fitness influencers, and healthcare providers.

The biohacking movement, propelled by charismatic figures like Dave Asprey and his Bulletproof brand, has packaged these concepts for a mass audience, promoting a suite of products and practices—from nootropic supplements and infrared saunas to ice baths—all promising optimized performance. This commercialization has significantly broadened the appeal of these practices but has also introduced a tension between evidence-based experimentation and market-driven hype. The core QS principle of n=1 (drawing conclusions based on one's own personal data) serves as a crucial counterbalance, encouraging individuals to verify for themselves whether a particular "hack" delivers a measurable benefit for their unique biology.

1.4.6 Critical Considerations: Privacy, Obsession, and the Meaning of Life

The ascent of the Quantified Self within biohacking is not without its significant criticisms and ethical dilemmas. The most pressing concern is data privacy. The intimate biometric data collected by wearables and apps constitutes a profoundly personal footprint. Questions arise over who owns this data, how it is stored, and how it might be used by technology companies, insurers, or employers.

Another major concern is the potential for these practices to foster anxiety and obsessive behaviors, a condition sometimes termed "orthosomnia"—the unhealthy preoccupation with achieving perfect sleep data. The relentless pursuit of optimization can paradoxically lead to diminished well-being, as individuals become disconnected from their bodily sensations and intrinsic motivations, instead living by the cold metrics on a screen. This raises a profound philosophical question: in the quest to quantify every aspect of our existence, do we risk reducing the rich, qualitative experience of life to a series of data points, potentially missing the forest for the trees?

1.5 THE PHILOSOPHIES BEHIND BIOHACKING: TRANSHUMANISM, SELF-OPTIMIZATION, AND LONGEVITY

The philosophies of transhumanism, self-optimization, and longevity form the intellectual backbone of the biohacking movement. Transhumanism provides the futuristic vision of human enhancement, self-optimization offers practical strategies for daily improvement, and longevity inspires the pursuit of extended and healthier lives. Together, these philosophies reveal that biohacking is not merely about techniques and tools but about deeper questions of identity, purpose, and the future of humanity. By blending aspirations for immediate performance, long-term vitality, and evolutionary transformation, biohacking reflects both the hopes and anxieties of a technological age. Whether embraced as visionary or criticized as reckless, these philosophies ensure that biohacking remains a powerful cultural and scientific phenomenon shaping the twenty-first century.

1.5.1 Transhumanism as the Foundation of Biohacking

Transhumanism is one of the most influential philosophies underpinning the biohacking movement. It is a worldview that advocates for using science and technology to transcend the natural limitations of the human body and mind. Transhumanists believe that evolution should not be left to chance but should be deliberately guided through technological intervention. For biohackers inspired by transhumanism, the body is not a fixed biological entity but a platform for continual improvement and enhancement. The philosophy of transhumanism envisions a future where humans integrate with artificial intelligence, genetic engineering, and nanotechnology to achieve radical transformations. Within biohacking, this manifests in practices such as neural implants, sensory augmentation, and the exploration of genetic modification. For example, grinders who implant RFID chips or magnets under their skin embody the transhumanist vision of human–machine integration. Transhumanism in biohacking is thus not limited to short-term health improvements but aspires to usher in a post-human future where individuals may possess cognitive, sensory, and physical abilities far beyond what evolution has provided.

Transhumanism and the Redefinition of Humanity

One of the core philosophical questions posed by transhumanism within biohacking is what it means to be human. By advocating for the augmentation of human biology, transhumanism challenges traditional notions of identity, mortality, and natural boundaries. For transhumanist biohackers, there is no inherent virtue in remaining biologically "natural." Instead, they argue that refusing to embrace technological enhancement is a form of stagnation.

This belief fuels the development of experimental technologies such as brain-computer interfaces, bionic limbs, and genetic therapies, which may one day erase the distinction between human and machine. Critics, however, caution that such philosophies could lead to new inequalities and ethical dilemmas, particularly if only certain individuals or societies gain access to enhancement technologies. Nevertheless, transhumanism remains a powerful force that shapes biohacking as both a practical pursuit and a cultural movement.

1.5.2 Self-Optimization as a Daily Philosophy

While transhumanism projects a long-term vision of human evolution, self-optimization reflects a more immediate and personal philosophy within biohacking. This approach focuses on maximizing physical, mental, and emotional performance in daily life through continuous experimentation. Self-optimization biohackers view the body as a system that can be tuned and adjusted much like a machine. Every variable—whether diet, sleep, exercise, or mental practice—can be measured, analyzed, and optimized to achieve peak performance.

In practice, this philosophy often involves meticulous tracking of biomarkers using wearable devices, adherence to specific dietary regimens such as intermittent fasting or ketogenic diets, and the use of nootropics to enhance cognitive performance. Meditation, mindfulness, and neurofeedback are also part of this approach, aimed at improving mental clarity and resilience. Unlike transhumanism, which is speculative and futuristic, self-optimization is pragmatic and grounded in measurable outcomes. Its appeal lies in its promise of immediate benefits such as improved energy, sharper focus, and better health.

Quantified Self Movement and Self-Optimization

Self-optimization within biohacking is closely linked to the quantified self movement, which emphasizes data-driven self-experimentation. Biohackers who embrace this philosophy often rely on devices that monitor heart rate variability, blood glucose levels, sleep cycles, and other physiological indicators. By analyzing this data, they adjust their routines to achieve maximum efficiency.

The underlying belief is that through constant measurement and refinement, individuals can push their bodies and minds to function at higher levels than would otherwise be possible. This philosophy reflects a cultural trend that values productivity, efficiency, and control, turning the pursuit of self-improvement into a technological lifestyle. However, critics argue that the obsession with self-optimization may lead to anxiety, over-monitoring, or the commodification of health. Nevertheless, it remains one of the most popular and accessible forms of biohacking philosophy.

1.5.3 Longevity as the Ultimate Goal

Another central philosophy driving biohacking is the pursuit of longevity. Unlike self-optimization, which focuses on immediate performance, the longevity philosophy seeks to extend not only lifespan but also healthspan, the number of years lived in good health. Longevity biohackers believe that aging is not an unavoidable fate but a biological process that can be slowed, halted, or even reversed through targeted interventions. This philosophy is closely tied to scientific research in fields such as gerontology, genetics, and regenerative medicine.

Longevity biohacking includes practices such as caloric restriction, fasting, supplementation with compounds like resveratrol or NAD+ boosters, and the use of advanced diagnostic tools for early disease detection. It also involves exploring cutting-edge interventions such as stem cell therapies and gene editing to repair or rejuvenate aging cells. The underlying conviction is that human life can be dramatically extended if science is applied proactively rather than reactively. For longevity-focused biohackers, the ultimate goal is not merely to live longer but to maintain vitality and functionality throughout extended lifespans.

Ethical Dimensions of Longevity Philosophy

The pursuit of longevity within biohacking raises important ethical and societal questions. If human lifespan could be significantly extended, what would be the implications for population growth, resource distribution, and social structures? Would access to life-extending technologies be limited to the wealthy, thereby creating new forms of inequality? Furthermore, the longevity philosophy challenges traditional cultural and religious views on mortality, forcing societies to rethink what it means to live a full life. While some critics see the quest for immortality as hubris, proponents argue that reducing suffering and extending health are natural extensions of medicine's purpose. Regardless of these debates, longevity remains one of the most ambitious and influential philosophies within biohacking.

1.5.4 Intersections of the Three Philosophies

Although transhumanism, self-optimization, and longevity can be viewed as distinct philosophies, they often intersect in the practice of biohacking. A biohacker who practices intermittent fasting may be motivated by self-optimization in terms of energy levels, by longevity in terms of cellular repair, and by transhumanist ideals of pushing human limits. Similarly, the use of nootropics can serve as a tool for immediate productivity while also reflecting the transhumanist vision of enhanced cognition. These overlaps demonstrate that biohacking is not guided by a single philosophy but by a constellation of beliefs that interact and reinforce one another. Together, they form a comprehensive framework that explains the motivations behind the movement and its appeal to diverse communities.

CHAPTER 2

Types of Biohacking

2.1 NUTRITIONAL BIOHACKING

2.1.1 Nootropics and Cognitive Enhancers

Nootropics and cognitive enhancers occupy a central role in the philosophy and practice of nutritional biohacking. They embody the drive to optimize brain function, gain control over mental states, and push beyond natural cognitive limits. Whether through natural supplements, synthetic smart drugs, or dietary interventions, biohackers use nootropics to experiment with the boundaries of human intelligence and resilience. While they hold great promise for productivity, creativity, and long-term brain health, they also raise questions about safety, fairness, and human identity in a performance-driven society. As research advances and cultural acceptance grows, nootropics will continue to shape not only the future of biohacking but also humanity's relationship with its most complex and powerful organ—the brain.

The future of nootropics and nutritional biohacking lies in personalization and technological integration. Advances in genomics and personalized medicine may soon allow individuals to tailor nootropic regimens based on their unique genetic profiles and metabolic responses. Artificial intelligence and data-driven health platforms could analyze biomarkers in real time to recommend specific compounds for optimal mental performance. Additionally, research into natural compounds and synthetic molecules will likely expand the range of available nootropics, offering more targeted and safer options. As the movement grows, the challenge will be balancing the desire for cognitive enhancement with ethical considerations, scientific rigor, and public safety.

Concept of Nootropics

The term "nootropic" was first coined in 1972 by Romanian chemist Corneliu Giurgea, who defined it as a compound that enhances learning and memory, protects the brain against injury, and has very few side effects. According to Giurgea's original definition, true nootropics not only improve cognition but also support long-term brain health. However, in modern usage, the term has broadened to encompass substances ranging from caffeine and herbal extracts to prescription medications like modafinil or racetams. Cognitive enhancers, on the other hand, may include more powerful substances that provide short-term mental performance boosts, sometimes at the expense of long-term safety. This blurred distinction shows the complexity of nutritional biohacking, where the line between beneficial supplementation and risky self-experimentation is not always clear.

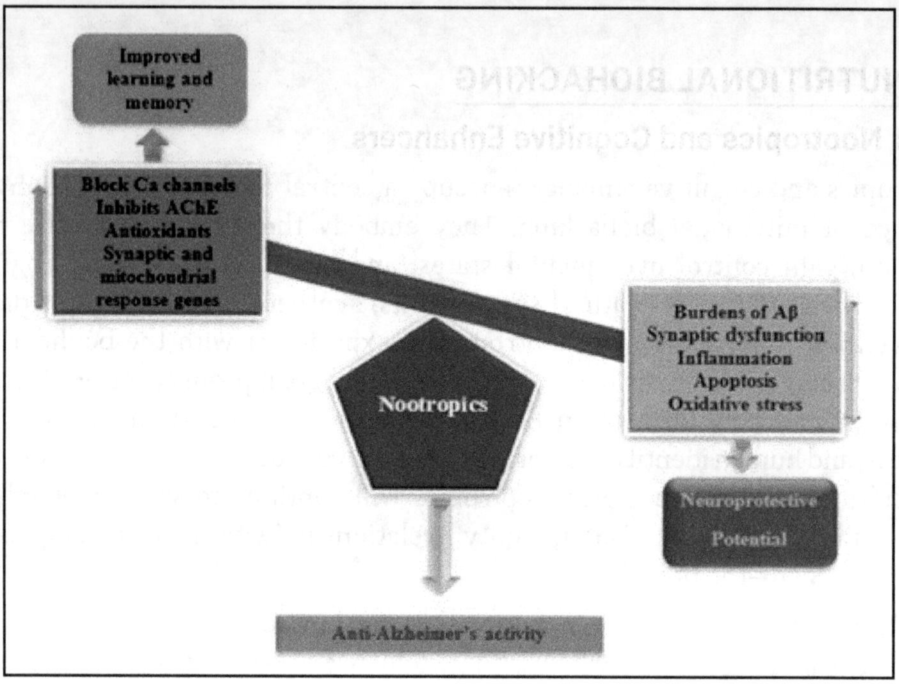

Fig: Nootropics

Nutritional Biohacking and the Brain

Nutritional biohacking is based on the principle that diet and supplementation can significantly influence brain function. The brain requires a steady supply of nutrients, neurotransmitters, and energy to operate at peak capacity. Nutritional deficiencies or imbalances can impair focus, memory, and emotional regulation, while targeted supplementation can enhance mental performance. For instance, omega-3 fatty acids from fish oil are known to support neuronal health, while vitamins such as B12 and folate play essential roles in neurotransmitter synthesis. Nootropics extend this principle by introducing substances that may fine-tune brain chemistry, optimize neurotransmission, or promote neuroplasticity. Through these interventions, biohackers aim to achieve sharper focus, quicker decision-making, and improved mental resilience in demanding environments.

Natural Nootropics and Dietary Enhancers

Many biohackers prefer to use natural nootropics derived from plants, herbs, and foods, viewing them as safer and more sustainable options. Caffeine, found in coffee and tea, is perhaps the most widely used nootropic, known for its ability to increase alertness and reduce fatigue. L-theanine, an amino acid present in green tea, is often combined with caffeine to promote calm focus without the jitters. Ginkgo biloba and Bacopa monnieri are traditional herbs that have been studied for their effects on memory and learning. Adaptogens like Rhodiola rosea and Ashwagandha are used to reduce stress and support mental endurance. Even common nutrients such as magnesium, vitamin D, and choline are considered nootropic in the sense that they enhance brain function when consumed in adequate amounts. These natural compounds embody the philosophy of nutritional biohacking, where food and supplements are used strategically to support cognitive health.

Synthetic Nootropics and Smart Drugs

Beyond natural enhancers, biohackers often experiment with synthetic nootropics designed to provide more powerful effects. The racetam family, beginning with piracetam, was one of the first groups of synthetic nootropics studied for cognitive enhancement. These compounds are believed to enhance communication between brain cells and improve memory consolidation. Modafinil, a prescription drug originally developed to treat narcolepsy, has gained popularity as a cognitive enhancer because it promotes wakefulness and focus without the stimulant crash associated with caffeine or amphetamines. Other synthetic nootropics include noopept, which may support memory and neuroprotection, and various designer compounds that target neurotransmitter systems in complex ways. While synthetic nootropics can provide substantial short-term benefits, they often carry greater risks, especially when used without medical supervision.

Mechanisms of Action in Cognitive Enhancement

The effectiveness of nootropics lies in their ability to influence brain chemistry and neurobiology. Some nootropics enhance blood flow to the brain, ensuring a greater supply of oxygen and nutrients for optimal performance. Others increase the availability of neurotransmitters such as acetylcholine, dopamine, and serotonin, which regulate mood, motivation, and memory. Compounds like Bacopa monnieri promote neuroplasticity, enhancing the brain's ability to adapt and form new connections. Adaptogens, on the other hand, modulate the body's stress response, reducing cortisol levels and protecting cognitive function under pressure. Stimulant-like nootropics, such as modafinil, work by increasing wakefulness-promoting neurotransmitters, thereby boosting attention and focus. By acting through multiple pathways, nootropics highlight the complexity of nutritional biohacking and its reliance on the intricate workings of the human brain.

Appeal of Nootropics in Modern Society

The rising popularity of nootropics reflects broader cultural values that emphasize productivity, efficiency, and cognitive performance. In highly competitive environments such as academia, entrepreneurship, and creative industries, the ability to think faster, remember more, and sustain attention is seen as a critical advantage. Nootropics promise to provide this edge by offering a chemical shortcut to mental optimization. For biohackers, the appeal goes beyond performance enhancement; it represents autonomy over one's cognitive destiny. By carefully choosing what substances to ingest, individuals feel empowered to control their mood, creativity, and intellectual output. This resonates with the broader ethos of biohacking, which rejects passivity and embraces active self-experimentation.

Risks and Ethical Concerns of Cognitive Biohacking

Despite their promise, nootropics and cognitive enhancers come with risks that cannot be ignored. The lack of long-term research on many synthetic compounds raises concerns about safety, dependency, and potential neurological damage. Even natural nootropics can cause side effects if used excessively or in combination with other substances. The use of prescription drugs like modafinil or Adderall for enhancement rather than treatment also raises ethical questions, particularly about fairness and coercion in academic or professional settings. If cognitive enhancers become normalized, individuals may feel pressured to use them simply to keep up with peers, creating a cycle of dependency. Moreover, unregulated online markets for nootropics make it difficult to ensure purity, dosage accuracy, and safety. These concerns highlight the fine line between responsible biohacking and reckless experimentation.

Nootropics and the Pursuit of Longevity

While nootropics are often associated with short-term boosts in performance, they are also linked to the philosophy of longevity within biohacking. Many compounds used as cognitive enhancers also support long-term brain health by reducing oxidative stress, promoting mitochondrial function, and protecting against neurodegeneration. For example, omega-3 fatty acids and antioxidants not only improve immediate cognition but also reduce the risk of age-related decline. Biohackers who focus on longevity view nootropics as a way to preserve mental sharpness and quality of life well into old age. In this sense, cognitive enhancement is not only about competitive advantage but also about sustaining vitality over a longer lifespan.

2.1.2 Ketogenic and Intermittent Fasting Protocols

The ketogenic diet and intermittent fasting represent two of the most effective and well-researched nutritional biohacking protocols available today. By leveraging the body's natural metabolic pathways, these approaches offer a science-backed method for enhancing energy, cognitive function, and longevity. However, their success depends on thoughtful implementation, personalization, and a commitment to overall health rather than rigid adherence to trends. As with any powerful tool, these protocols must be used wisely, with attention to individual needs and potential risks. For those willing to experiment, track, and adjust, ketogenic and intermittent fasting protocols can be transformative, unlocking new levels of physical and mental performance. Ultimately, the goal of nutritional biohacking is not perfection but rather a deeper understanding of one's own biology and the ability to make informed, intentional choices that support long-term health and vitality.

As research in metabolic health and longevity continues to advance, the future of nutritional biohacking looks increasingly promising. Emerging technologies, such as AI-driven personalized nutrition platforms and advanced biomarker tracking, are making it easier for individuals to optimize their diets and fasting protocols based on real-time biological feedback.

The integration of genetic testing and microbiome analysis may further refine these approaches, allowing for highly individualized recommendations that account for unique genetic predispositions and gut health. Additionally, the growing body of research on the interplay between diet, fasting, and cellular aging suggests that these protocols may play a significant role in extending healthspan and delaying age-related diseases. However, as the field evolves, it will be essential to maintain a balanced perspective, recognizing that while these tools can be powerful, they are not one-size-fits-all solutions. The future of nutritional biohacking will likely emphasize a holistic approach, where metabolic optimization is integrated with mental health, physical activity, and sustainable lifestyle practices to achieve true, long-lasting well-being.

The Science behind the Ketogenic Diet

The ketogenic diet is founded on the principle of metabolic switching, where the body transitions from relying on glucose as its primary energy source to burning fat-derived ketones. This shift occurs when carbohydrate intake is drastically reduced, typically to less than 50 grams per day, depleting glycogen stores and prompting the liver to produce ketones from fatty acids. Ketones are a highly efficient fuel source for the brain and muscles, often leading to improved mental clarity, sustained energy, and reduced inflammation. Research has shown that ketosis can enhance mitochondrial function, increase the production of brain-derived neurotrophic factor (BDNF), and reduce oxidative stress, all of which contribute to neuroprotection and cognitive enhancement. Additionally, the ketogenic diet has been studied for its potential therapeutic benefits in conditions such as epilepsy, type 2 diabetes, and neurodegenerative diseases. However, the diet is not without challenges, as the initial adaptation phase, often called the "keto flu," can cause fatigue, headaches, and irritability as the body adjusts to burning fat for fuel. Long-term adherence also requires careful attention to nutrient density, as deficiencies in electrolytes, vitamins, and minerals can arise if the diet is not properly balanced.

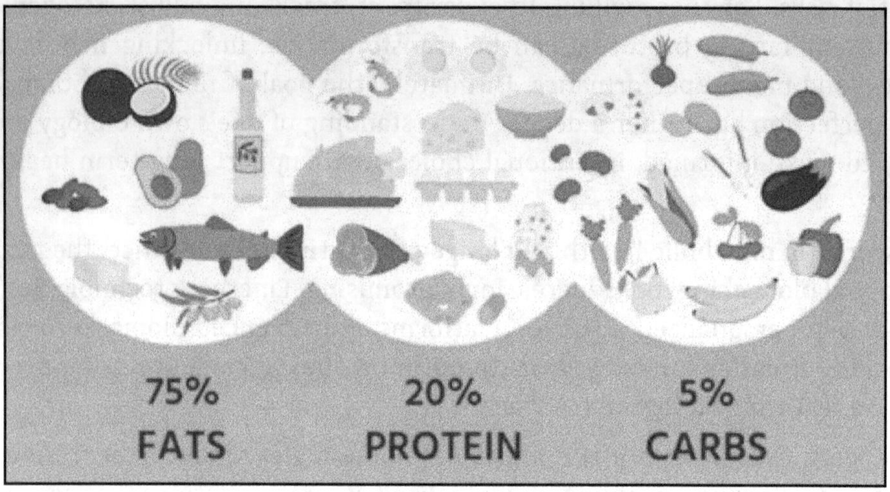

Fig: Ketosis Diet

Intermittent Fasting: A Metabolic Reset

Intermittent fasting is another powerful biohacking tool that manipulates eating patterns to optimize metabolic health. Unlike traditional calorie restriction, which focuses on reducing overall food intake, intermittent fasting emphasizes the timing of meals, creating distinct periods of feeding and fasting. Common protocols include the 16:8 method, where individuals fast for 16 hours and eat within an 8-hour window, and the 5:2 approach, which involves eating normally for five days and significantly reducing calorie intake for two non-consecutive days. The primary benefit of intermittent fasting lies in its ability to enhance metabolic flexibility, improve insulin sensitivity, and promote autophagy, the body's natural process of cellular cleanup and repair. Fasting also triggers the release of human growth hormone (HGH), which supports muscle preservation and fat loss, and reduces levels of insulin-like growth factor 1 (IGF-1), a marker associated with aging and age-related diseases. Beyond its metabolic effects, intermittent fasting has been shown to support brain health by increasing BDNF and reducing neuroinflammation, potentially lowering the risk of cognitive decline. However, fasting is not universally suitable, as individuals with certain medical conditions, such as diabetes or a history of eating disorders, may experience adverse effects and should approach fasting with caution.

Combining Ketogenic and Intermittent Fasting Protocols

When used together, the ketogenic diet and intermittent fasting can create a synergistic effect, amplifying the metabolic benefits of each approach. The ketogenic diet primes the body to burn fat efficiently, while intermittent fasting enhances this process by extending the period during which the body relies on fat stores for energy. This combination can lead to accelerated fat loss, improved mental clarity, and enhanced cellular repair mechanisms.

Additionally, fasting in a ketogenic state may further elevate ketone levels, providing a more stable and sustained energy source for the brain and body. However, combining these protocols requires careful planning to avoid potential pitfalls, such as excessive calorie restriction, electrolyte imbalances, or nutrient deficiencies. It is essential to prioritize nutrient-dense foods, stay hydrated, and monitor electrolyte levels, particularly sodium, potassium, and magnesium, which can become depleted during fasting and ketosis. Some individuals may also experience increased fatigue or difficulty concentrating during the initial adaptation phase, necessitating gradual implementation and close attention to individual tolerance.

Practical Implementation and Personalization

Successfully adopting ketogenic and intermittent fasting protocols requires more than just following a rigid set of rules; it demands personalization based on individual metabolism, lifestyle, and health goals. For those new to ketogenic eating, a gradual reduction in carbohydrate intake can help ease the transition into ketosis, minimizing the discomfort of the adaptation phase. Similarly, intermittent fasting should be introduced progressively, starting with shorter fasting windows and gradually extending them as the body adapts.

Tracking tools such as continuous glucose monitors (CGMs), ketone meters, and fitness wearables can provide valuable feedback, allowing individuals to fine-tune their approach based on real-time data. It is also important to consider the timing of meals and macronutrient distribution, as eating larger meals during the feeding window can help maintain energy levels and support metabolic health. Additionally, incorporating strategies such as time-restricted eating, where meals are consumed earlier in the day to align with circadian rhythms, can further enhance the benefits of fasting. Ultimately, the key to sustainable success lies in experimentation and adjustment, as each person's response to these protocols can vary widely based on genetics, activity levels, and overall health status.

Potential Risks and Considerations

While ketogenic and intermittent fasting protocols offer numerous benefits, they are not without potential risks, particularly when implemented without proper guidance or individualization. One of the primary concerns with the ketogenic diet is the risk of nutrient deficiencies, as the elimination of certain food groups can lead to inadequate intake of essential vitamins and minerals. Additionally, some individuals may experience digestive issues, such as constipation, due to the low fiber content of many ketogenic diets. Intermittent fasting, particularly when taken to extremes, can lead to excessive stress on the body, disrupting hormonal balance and potentially exacerbating conditions such as adrenal fatigue or thyroid dysfunction. There is also the risk of developing an unhealthy relationship with food, where rigid fasting schedules or restrictive eating patterns lead to disordered eating behaviors. It is crucial for individuals to listen to their bodies, seek professional guidance when necessary, and prioritize overall well-being over strict adherence to protocols. Those with pre-existing medical conditions, such as diabetes, kidney disease, or a history of eating disorders, should consult with a healthcare provider before adopting these strategies to ensure they are implemented safely and appropriately.

2.1.3 Supplements and Functional Foods

Defining the Toolkit: Supplements versus Functional Foods

Within the framework of nutritional biohacking, the arsenal is divided into two primary, yet often overlapping, categories: dietary supplements and functional foods. Supplements are concentrated sources of nutrients or other substances with a nutritional or physiological effect, designed to be taken in addition to the normal diet. They are typically delivered in pill, powder, or liquid form and include a vast range of products from foundational vitamins and minerals to more specialized compounds like adaptogenic herbs, nootropics, probiotics, and exogenous ketones.

Their purpose is to provide a potent, measured dose of a specific compound that may be difficult to obtain in sufficient quantities from food alone. Functional foods, on the other hand, are whole foods or fortified foods that provide a health benefit beyond basic nutrition. They are consumed as part of a regular diet but are chosen specifically for their additional bioactive components. Examples include turmeric for its anti-inflammatory curcumin, fatty fish for omega-3 fatty acids, fermented foods like kimchi and kefir for their probiotics, and dark leafy greens for their array of phytonutrients and minerals. The distinction blurs with products like protein powders or greens powders, which are supplemental in form but are often used to functionalize meals and shakes.

Philosophical Foundation: From General Health to Targeted Intervention

Nutritional biohacking is fundamentally driven by a shift from a generalized, one-size-fits-all health recommendation to a personalized, data-informed strategy. Traditional nutritional advice often operates on population-level guidelines aimed at preventing deficiency diseases and promoting general well-being. While valuable, this approach does not account for individual genetic variability, gut microbiome composition, metabolic health, and specific performance goals. The biohacker's philosophy is rooted in the concept of biochemical individuality. It asks not "What is a healthy diet?" but rather "What does my body need to perform at its optimal level?" This necessitates a process of experimentation and self-quantification. A biohacker might use blood tests to identify micronutrient deficiencies, continuous glucose monitors to understand personal carbohydrate tolerance, or sleep trackers to correlate dietary choices with recovery quality. The resulting data informs a highly personalized nutritional protocol where specific supplements and functional foods are deployed to correct imbalances, enhance strengths, and achieve precise objectives, whether that is sharpening mental focus, reducing systemic inflammation, or improving metabolic flexibility.

Key Domains of Intervention: How Nutritional Biohacking Manifests

The application of supplements and functional foods within biohacking spans several key domains of human performance. In the realm of cognitive enhancement, biohackers utilize nootropic supplements like lion's mane mushroom for nerve growth factor support, phosphatidylserine for stress resilience, and functional foods such as blueberries and walnuts rich in antioxidants and healthy fats to protect and nourish neural tissues. For physical performance and recovery, the toolkit includes branched-chain amino acids (BCAAs) to reduce muscle soreness, creatine monohydrate for strength and power output, and anti-inflammatory functional foods like tart cherries and ginger to modulate exercise-induced inflammation. A major focus is also on gut health, recognizing the gut as the foundation of overall well-being. Here, interventions include probiotic supplements to reseed beneficial bacteria, prebiotic fibers from foods like garlic and asparagus to feed them, and bone broth for its gut-healing amino acids like glutamine. Furthermore, the pursuit of longevity has brought attention to supplements like NAD+ precursors (e.g., NMN, NR) for cellular energy and repair, resveratrol for activating longevity pathways, and a diet rich in polyphenol-dense functional foods to combat oxidative stress.

Critical Considerations and Responsible Practice

The world of nutritional biohacking, while empowering, is not without its significant challenges and risks. The supplement industry is notoriously under-regulated, leading to issues with product purity, inaccurate labeling, and potentially harmful contaminants. The mantra of "self-experimentation" can also be dangerous without proper knowledge, as megadoses of certain fat-soluble vitamins or herbs can lead to toxicity, adverse effects, and harmful interactions with medications. The very premise of optimization can sometimes veer into orthorexic tendencies, an unhealthy obsession with eating "perfectly." Therefore, a responsible approach to nutritional biohacking is paramount. This involves prioritizing a whole-food, nutrient-dense diet as the foundational base, using supplements only to address identified and specific needs. Sourcing products from reputable, third-party tested companies and, crucially, consulting with knowledgeable healthcare professionals—such as registered dietitians or functional medicine doctors—can help navigate the complexities and ensure that interventions are safe, effective, and truly beneficial rather than merely following trends.

2.2 LIFESTYLE AND BEHAVIORAL HACKS

2.2.1 Sleep Optimization

Sleep is one of the most fundamental aspects of human health, yet it is also one of the most neglected in modern society. With the rise of demanding lifestyles, constant digital distractions, and high levels of stress, sleep quality has deteriorated for many people. Sleep optimization has therefore emerged as a growing field within health science and biohacking, focusing on strategies to improve the duration, depth, and restorative quality of sleep. Rather than simply emphasizing the number of hours spent in bed, sleep optimization looks at how sleep can be fine-tuned to support physical recovery, mental performance, and long-term well-being. By understanding the biological mechanisms of sleep and implementing intentional practices, individuals can unlock the full potential of this natural process.

Science of Sleep

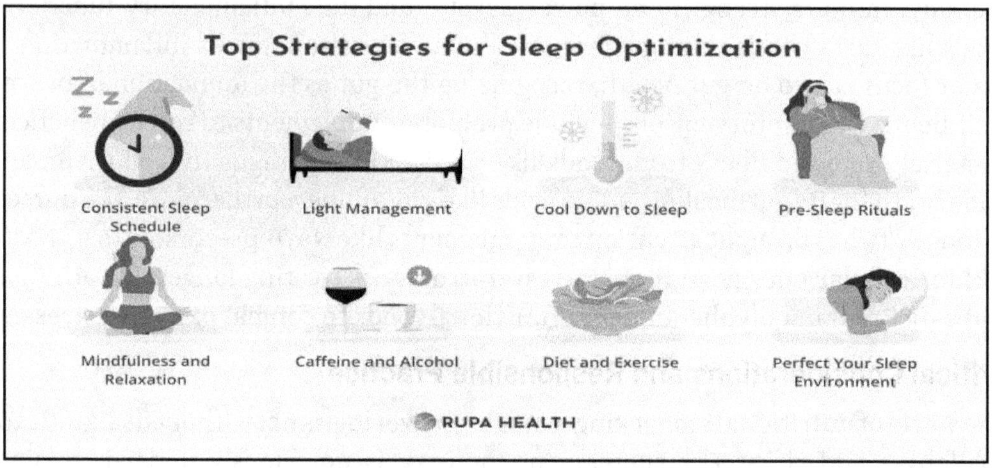

Fig: Strategies for Sleep Optimization

Sleep is not a passive state but an active biological process essential for restoration, memory consolidation, and metabolic regulation. It occurs in cycles that alternate between non-rapid eye movement (NREM) sleep and rapid eye movement (REM) sleep. Within NREM sleep, deep slow-wave stages are critical for physical recovery, hormone regulation, and immune system function. REM sleep, on the other hand, plays a central role in emotional processing and learning. The circadian rhythm, often referred to as the body's internal clock, regulates the timing of sleep and wakefulness through the release of hormones such as melatonin and cortisol. Sleep optimization therefore requires not only ensuring adequate total sleep but also protecting the quality of these cycles and aligning them with the natural rhythm of the body.

Modern Challenges to Sleep Quality

Despite its importance, sleep is increasingly disrupted in contemporary society. The widespread use of screens and artificial lighting at night interferes with melatonin production, making it harder for the body to prepare for rest. Stress and anxiety, fueled by fast-paced work environments and constant connectivity, often result in insomnia or restless sleep.

Irregular schedules, such as those experienced by shift workers, disrupt the circadian rhythm and lead to long-term health consequences. Additionally, environmental factors such as noise, temperature, and exposure to blue light contribute to poor sleep quality. As a result, many individuals struggle not only with falling asleep but also with maintaining consistent, restorative rest throughout the night. Sleep optimization aims to address these challenges by creating conditions that support the body's natural sleep processes.

Sleep Hygiene as a Foundation

One of the most effective approaches to sleep optimization is practicing good sleep hygiene. This refers to the set of behaviors and environmental factors that create an optimal context for sleep. Regularity is key, as going to bed and waking up at consistent times helps reinforce the circadian rhythm. The sleep environment should be quiet, cool, and dark, allowing the body to enter deep sleep stages without interruption. Reducing screen exposure at least an hour before bedtime is also crucial, as the blue light from devices suppresses melatonin. Relaxation techniques such as reading, meditation, or gentle stretching signal to the body that it is time to wind down. By establishing these practices, individuals create a supportive framework for high-quality sleep on a nightly basis.

Role of Nutrition in Sleep Optimization

Nutrition plays a significant role in determining sleep quality. Heavy meals, caffeine, and alcohol close to bedtime often disrupt sleep cycles and prevent deep rest. Conversely, certain nutrients support sleep by influencing neurotransmitters involved in relaxation. Foods rich in magnesium, such as nuts and leafy greens, help calm the nervous system. Tryptophan, an amino acid found in foods like turkey and dairy, promotes the production of serotonin and melatonin, both essential for sleep regulation. Herbal teas containing chamomile or valerian root are also widely used as natural sleep aids. Sleep optimization through nutrition therefore involves not only avoiding disruptive substances but also including foods that naturally support rest and recovery.

Technology and Sleep Tracking

Advancements in wearable technology have provided new tools for sleep optimization by allowing individuals to monitor their sleep patterns in detail. Devices such as smartwatches and sleep trackers measure metrics including sleep stages, heart rate variability, and nighttime movement. This data-driven approach, often referred to as part of the quantified self-movement, enables individuals to experiment with different habits and observe how they affect sleep quality. For example, one might compare how caffeine consumption, exercise timing, or room temperature influence the proportion of deep and REM sleep. While sleep trackers are not always perfectly accurate, they provide valuable insights that guide personalized approaches to sleep optimization.

Psychological Strategies for Better Sleep

Mental and emotional states have a profound effect on the ability to fall and stay asleep. Chronic stress activates the sympathetic nervous system, making it difficult for the body to transition into rest. Psychological strategies such as mindfulness meditation, journaling, and controlled breathing techniques help calm the mind and prepare it for sleep. Cognitive behavioral therapy for insomnia (CBT-I) has been shown to be one of the most effective treatments for chronic sleep problems, as it addresses thought patterns and behaviors that interfere with rest. Visualization practices, gratitude exercises, and progressive muscle relaxation are also powerful tools for quieting the mind before bed. Incorporating these strategies into a nightly routine ensures that sleep optimization is not limited to physical habits but also includes mental well-being.

Sleep Optimization for Physical and Mental Performance

Optimizing sleep is not only about preventing fatigue but also about enhancing performance in daily life. During deep sleep, the body releases growth hormone, which is essential for muscle repair and physical recovery. This makes sleep critical for athletes and anyone engaging in regular physical activity. REM sleep, on the other hand, supports creativity, problem-solving, and emotional balance. By improving sleep quality, individuals experience sharper focus, improved memory, and greater resilience in managing stress. Sleep optimization therefore serves as a foundation for both physical vitality and mental excellence, making it a core aspect of long-term self-improvement.

Connection between Sleep and Longevity

Research has consistently shown that sleep is one of the most important predictors of longevity and long-term health. Poor sleep is associated with higher risks of cardiovascular disease, diabetes, obesity, and neurodegenerative conditions such as Alzheimer's disease. Chronic sleep deprivation also weakens the immune system, leaving individuals more vulnerable to infections and illnesses. By contrast, consistent, high-quality sleep supports cellular repair, regulates metabolism, and protects brain function. Sleep optimization can therefore be seen as a form of preventive medicine, ensuring not only better performance in the present but also greater vitality and health in later years. For biohackers focused on longevity, sleep is considered one of the most powerful natural interventions available.

2.2.2 Meditation and Breathwork Techniques

Internal Technology: Meditation and Breathwork

In the expansive toolkit of biohacking, where external devices and supplements often claim the spotlight, a more ancient and profound technology resides within: the conscious control of one's own mind and breath. Meditation and breathwork represent the foundational pillars of internal biohacking, offering a zero-cost, universally accessible means to directly influence the nervous system, reshape brain architecture, and optimize physiological function.

Unlike interventions that add something to the body, these practices work by subtracting—reducing the noise of stress, the chatter of the mind, and the unconscious patterns of breathing that keep the body in a state of high alert. They are not merely relaxation techniques but are powerful modalities for enhancing cognitive clarity, emotional regulation, and physical resilience, providing a master key to systemic well-being.

Science of Stillness: Understanding Meditation

Meditation, at its core, is the systematic training of attention and awareness. It is a deliberate practice that cultivates a state of mindful presence, allowing the practitioner to observe the contents of their consciousness—thoughts, emotions, sensations—without becoming entangled in them. Neuroscientific research has validated what contemplative traditions have known for millennia: consistent meditation physically alters the brain.

Through neuroplasticity, the brain rewires itself in response to this repeated mental exercise. Studies using MRI scans have shown that meditation can increase gray matter density in regions associated with learning, memory, and emotional regulation, such as the prefrontal cortex and hippocampus, while decreasing gray matter in the amygdala, the brain's fear center. This neurological shift translates into tangible benefits: enhanced focus and concentration, reduced reactivity to stress, improved emotional intelligence, and a greater overall sense of calm. For the biohacker, meditation is not a spiritual abstraction but a reproducible protocol for upgrading the brain's hardware and software, leading to improved decision-making, creativity, and mental performance under pressure.

Fig: Meditation in 12 Steps

Art of Conscious Respiration: Exploring Breathwork

If meditation is the training of the mind, breathwork is the direct modulation of the autonomic nervous system through the conscious control of breathing patterns. Breathing is unique among autonomic functions; it operates automatically but can also be willingly controlled, making it a bridge between the conscious and unconscious realms of physiology. By altering the rhythm, depth, and pace of the breath, one can swiftly induce distinct physiological states. Slow, deep, diaphragmatic breathing—often at a pace of five or six breaths per minute—activates the parasympathetic nervous system, triggering the body's "rest and digest" response. This lowers heart rate, reduces blood pressure, and dampens the production of stress hormones like cortisol. In contrast, rapid and rhythmic breathing techniques, such as those found in practices like Holotropic

Breathwork or the Wim Hof Method, can temporarily induce a controlled stress response, sympathetic activation, and altered states of consciousness. This form of controlled hypoxia can train the body to become more resilient to stress, reduce inflammation, and release trapped emotional energy. Breathwork is thus a potent and immediate tool for biohackers to manage energy levels, control stress in real-time, and explore the interplay between physiology and consciousness.

Practical Integration: Techniques for the Modern Biohacker

The integration of these practices into a modern biohacking regimen does not require retreating to a monastery; it demands consistent, daily micro-sessions. A foundational practice is mindfulness meditation, which involves sitting quietly and focusing attention on the sensation of the breath. When the mind inevitably wanders, the practice is simply to notice the distraction and gently return focus to the breath without judgment. This strengthens the neural circuits of attention.

For a more technology-assisted approach, biohackers often use EEG headbands that provide real-time feedback on brainwave states, helping users learn to achieve and sustain calm, focused gamma or alpha waves. For breathwork, a simple starting protocol is the 4-7-8 technique: inhale quietly through the nose for a count of four, hold the breath for a count of seven, and exhale completely through the mouth for a count of eight. This pattern is a powerful nervous system sedative. More advanced practitioners might engage in cyclical hyperventilation followed by breath retention, as in the Wim Hof Method, to boost adrenaline, alkalize the body, and strengthen immune response. The key is consistency and self-experimentation, finding the techniques and durations that yield the most significant subjective and objective benefits for the individual.

Synergistic Effects on Health and Performance

The combined practice of meditation and breathwork creates a powerful synergy that amplifies their individual benefits, impacting nearly every domain of human performance. Cognitively, they act as a reset button for a fatigued brain, enhancing neurogenesis and improving cognitive flexibility, which allows for faster learning and better problem-solving. Physiologically, the reduction in chronic stress levels lowers systemic inflammation, a root cause of numerous modern diseases, and improves cardiovascular health.

For athletic performance, breathwork increases carbon dioxide tolerance, allowing for more efficient oxygen utilization and improved endurance, while the mental resilience cultivated through meditation enhances focus and pain management during competition. Furthermore, these practices significantly impact emotional health by creating space between a stimulus and one's reaction, allowing for more thoughtful responses rather than impulsive ones. This emotional regulation is perhaps one of the most valuable biohacks for navigating the complexities of modern life and maintaining high performance in demanding environments.

Beyond Optimization: The Deeper Dimension

While often adopted for their performance-enhancing benefits, meditation and breathwork ultimately lead the practitioner beyond mere optimization into a deeper dimension of self-inquiry. The constant pursuit of external biohacks—better supplements, smarter drugs, newer devices—can sometimes foster a sense that something is lacking that needs to be fixed.

The internal practices of meditation and breathwork counter this by cultivating a sense of wholeness and presence. They invite an exploration of consciousness itself, moving the focus from having a better-performing body to being a more aware and conscious individual. This is the ultimate hack: realizing that the state of one's nervous system and the quality of one's awareness are the ultimate determinants of life experience. In a world of endless external stimulation and demand, these ancient internal technologies provide the necessary counterbalance, offering a path to not just a higher-performing life, but a wiser, more centered, and more fulfilling one.

2.2.3 Cold Exposure and Heat Therapy

Cold exposure and heat therapy represent two of the most accessible and scientifically supported tools in the biohacker's arsenal. By harnessing the body's natural responses to temperature extremes, these practices offer a powerful means of enhancing physical recovery, cognitive function, and metabolic health. However, their effectiveness depends on thoughtful implementation, personalization, and a commitment to safety and moderation. For those willing to experiment and adapt, thermal biohacking can be a transformative addition to a holistic wellness routine, unlocking new levels of resilience and performance. Ultimately, the goal is not merely to endure temperature stress but to leverage it as a tool for mastering the body's adaptive potential, achieving a state of optimal health and vitality.

As research continues to uncover the mechanisms behind cold exposure and heat therapy, their applications in health optimization and performance enhancement are likely to expand. Advances in wearable technology, such as continuous temperature monitoring and real-time biomarker tracking, may allow for more precise and personalized thermal protocols. The integration of AI and machine learning could further refine these practices, enabling individuals to optimize their thermal exposure based on real-time physiological feedback. Additionally, the growing interest in longevity and anti-aging research suggests that thermal biohacking may play a significant role in promoting cellular health and resilience. However, as these practices become more mainstream, it will be essential to balance innovation with safety, ensuring that the pursuit of optimization does not come at the expense of long-term health. The future of thermal biohacking will likely emphasize a holistic approach, where temperature interventions are integrated with other wellness practices, such as nutrition, sleep optimization, and stress management, to achieve comprehensive and sustainable health benefits.

Science of Cold Exposure

Cold exposure, whether through ice baths, cold showers, or cryotherapy, works by subjecting the body to temperatures significantly lower than its comfort zone, typically between 50°F (10°C) and freezing. This thermal stress activates a cascade of physiological responses designed to protect core body temperature and maintain homeostasis. One of the most immediate effects is vasoconstriction, where blood vessels narrow to preserve heat, followed by vasodilation upon rewarming, which enhances circulation and lymphatic drainage.

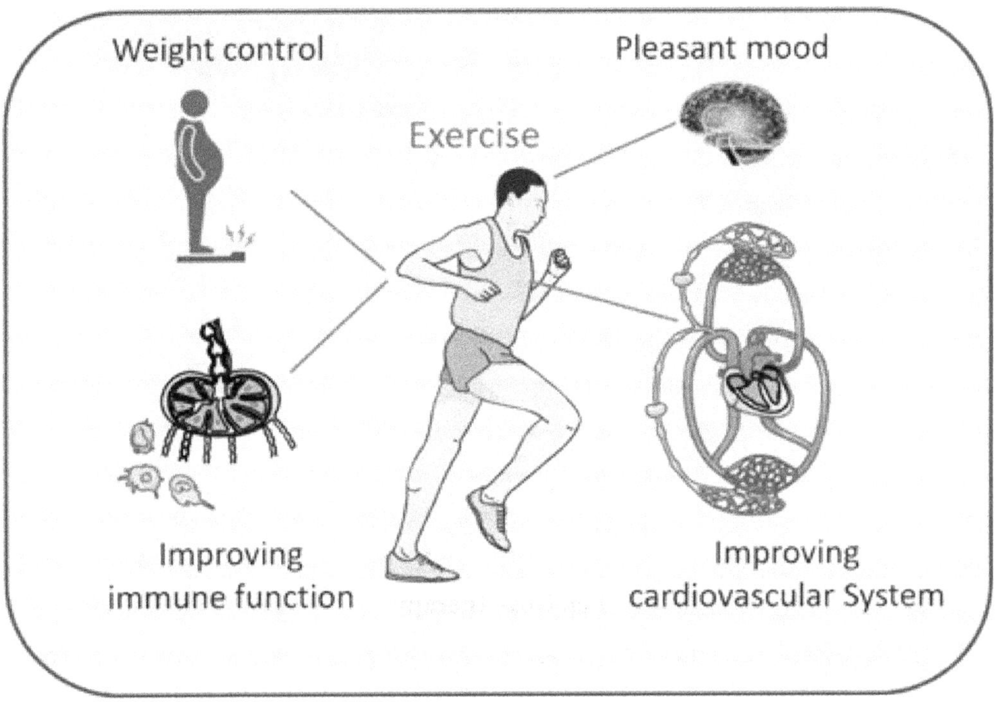

Fig: Cold Exposure

This process helps reduce muscle soreness and inflammation, making cold therapy a staple in athletic recovery protocols. Beyond its physical benefits, cold exposure has been shown to stimulate the release of norepinephrine, a neurotransmitter that enhances focus, mood, and energy levels. Regular cold exposure may also increase brown adipose tissue (BAT) activity, a type of fat that generates heat and burns calories, potentially aiding in metabolic health and weight management. Additionally, the practice has been linked to improved immune function, as the body's stress response to cold can enhance white blood cell activity. However, the intensity and duration of cold exposure must be carefully managed to avoid risks such as hypothermia or excessive stress on the cardiovascular system, particularly for those with pre-existing health conditions.

Heat Therapy and Its Physiological Benefits

On the opposite end of the thermal spectrum, heat therapy—through saunas, hot baths, or heated environments—induces a different set of adaptive responses. When exposed to high temperatures, typically between 150°F (65°C) and 200°F (93°C), the body initiates thermoregulatory mechanisms to cool itself, primarily through sweating and increased blood flow to the skin. This process not only aids in detoxification through sweat but also promotes cardiovascular health by mimicking the effects of moderate exercise.

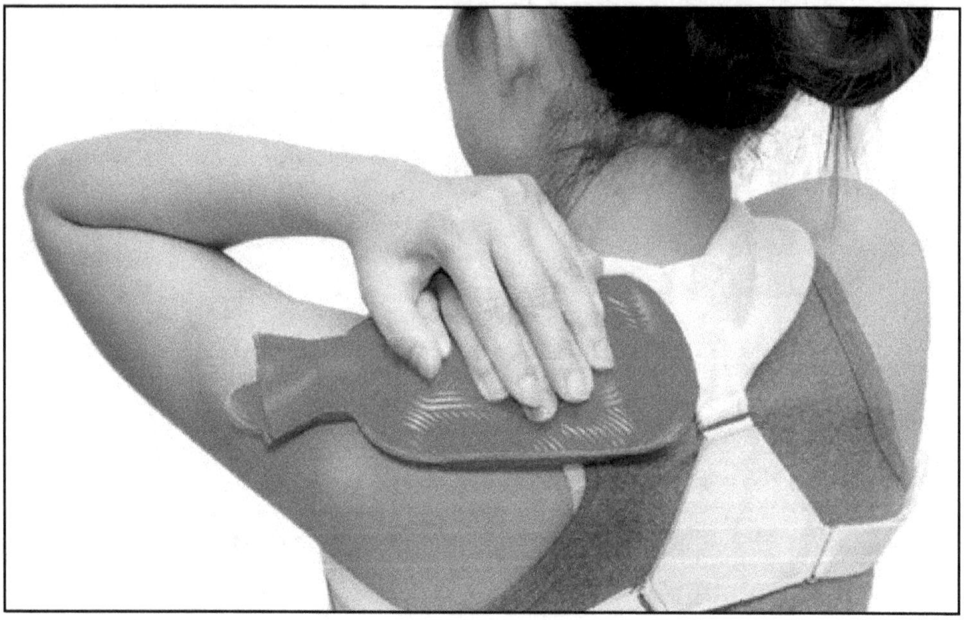

Fig: Heat Therapy

Heat exposure has been shown to increase heart rate and cardiac output, improving vascular function and reducing blood pressure over time. Additionally, heat therapy stimulates the production of heat shock proteins (HSPs), which play a crucial role in cellular repair and protection against stress, potentially enhancing longevity and resilience to disease. The practice has also been associated with reduced muscle tension, improved joint mobility, and enhanced recovery from physical exertion. Furthermore, regular heat exposure may support mental health by increasing endorphin release, which can alleviate stress and promote relaxation. However, prolonged or excessive heat exposure can lead to dehydration, dizziness, or even heat stroke, necessitating careful monitoring of duration and hydration levels.

Cold Exposure vs. Heat Therapy: Complementary Effects

While cold exposure and heat therapy may seem like opposing practices, they can be strategically combined to create a synergistic effect on health and performance. Cold therapy is particularly effective for acute recovery, reducing inflammation, and sharpening mental focus, making it ideal for post-workout recovery or morning routines to enhance alertness. Heat therapy, on the other hand, excels in promoting relaxation, muscle recovery, and cardiovascular health, making it a valuable tool for evening wind-down routines or pre-sleep relaxation.

Some biohackers alternate between the two, a practice known as contrast therapy, which leverages the benefits of both temperature extremes to enhance circulation, reduce muscle soreness, and improve overall recovery. This approach is often used in athletic training and rehabilitation, where the alternating vasoconstriction and vasodilation effects can accelerate healing and reduce delayed onset muscle soreness (DOMS). The key to maximizing the benefits of both modalities lies in timing and individualization, as the body's response to temperature stress can vary based on factors such as fitness level, health status, and personal tolerance.

Practical Applications and Protocols

Implementing cold exposure and heat therapy effectively requires an understanding of the appropriate protocols and safety considerations. For cold exposure, beginners often start with cold showers, gradually progressing to ice baths or whole-body cryotherapy sessions. A common protocol involves immersing the body in cold water (around 50°F or 10°C) for 10 to 15 minutes, though even shorter durations of 2 to 5 minutes can yield benefits. The Wim Hof Method, which combines cold exposure with specific breathing techniques, has gained popularity for its ability to enhance stress resilience and immune function.

For heat therapy, traditional saunas (dry heat) and infrared saunas (which penetrate deeper into tissues) are the most common methods. Sessions typically last between 15 to 30 minutes at temperatures ranging from 150°F to 190°F (65°C to 88°C), with hydration being a critical factor to prevent overheating. Some individuals integrate both practices into their daily or weekly routines, using cold exposure in the morning for energy and focus and heat therapy in the evening for relaxation and recovery. The frequency and intensity of these sessions should be adjusted based on individual response, with careful attention to signs of overexposure, such as excessive fatigue, dizziness, or prolonged discomfort.

Potential Risks and Considerations

Despite their benefits, cold exposure and heat therapy are not without risks, particularly when practiced without proper guidance or individualization. Cold exposure, especially in extreme forms like ice baths or cryotherapy, can pose risks such as hypothermia, frostbite, or cardiovascular strain, particularly for individuals with heart conditions or poor circulation. It is crucial to gradually acclimate the body to cold stress and to avoid prolonged exposure without proper monitoring.

Similarly, heat therapy can lead to dehydration, overheating, or even heat stroke if sessions are too long or temperatures too high. Individuals with cardiovascular issues, low blood pressure, or certain medications that affect thermoregulation should exercise caution and consult a healthcare provider before engaging in intense heat exposure. Additionally, both practices can be physically and mentally demanding, requiring a degree of discipline and self-awareness to avoid excessive stress or burnout. As with any biohacking practice, the key to safe and effective thermal therapy lies in moderation, gradual adaptation, and attentiveness to the body's signals.

2.3 DIGITAL AND WEARABLE BIOHACKING

2.3.1 Fitness and Health Trackers

Evolution of Fitness and Health Tracking

The history of fitness and health trackers can be traced back to simple pedometers, which were first developed centuries ago to count steps. These basic devices laid the foundation for modern wearables by highlighting the motivational power of tracking physical activity. With the rise of digital technology and sensors, fitness trackers evolved into sophisticated tools capable of monitoring a wide range of biological markers. Companies like Fitbit, Garmin, and Apple pioneered the wearable movement, creating sleek devices that combined convenience with accuracy.

Over time, health trackers have expanded beyond fitness to include medical applications such as monitoring blood oxygen saturation, detecting irregular heartbeats, and even predicting health risks. This evolution illustrates the growing intersection between consumer technology and healthcare.

Core Features of Fitness Trackers

Modern fitness and health trackers typically come equipped with a suite of sensors designed to capture different aspects of health. Step counting remains a foundational feature, motivating users to reach daily activity goals. Heart rate monitoring allows individuals to assess workout intensity, track resting heart rate, and detect fluctuations that may signal stress or fatigue. Sleep tracking analyzes patterns of light, deep, and REM sleep, offering insights into rest quality. Many devices also provide calorie expenditure estimates, helping users balance activity with nutrition. More advanced models include features such as GPS for tracking outdoor activities, blood oxygen sensors for measuring oxygen saturation, and electrocardiogram (ECG) capabilities for monitoring heart rhythm. Together, these features create a holistic picture of health that can guide users in making informed lifestyle choices.

Role of Data in Personal Health

The data collected by fitness and health trackers is at the core of their value. By providing objective feedback, these devices encourage individuals to reflect on their habits and make measurable improvements. For example, someone who consistently falls short of their step goal may be motivated to walk more throughout the day.

Similarly, users who notice poor sleep quality may adjust their bedtime routines to improve rest. This shift toward data-driven health is empowering, as it gives individuals control over their wellness in a way that was previously accessible only in clinical settings. Data also enables long-term tracking, allowing people to see progress over weeks, months, and even years, reinforcing consistency and accountability in health behaviors.

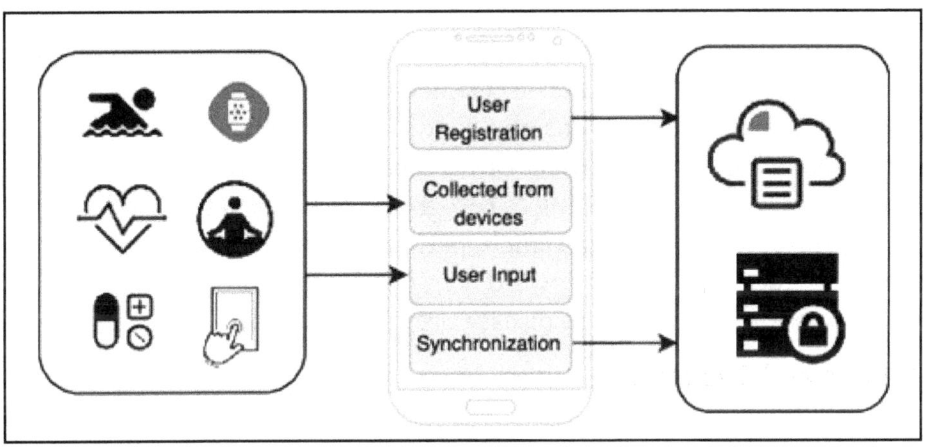

Fig: Personal Health Data Collection

Psychological Impact and Motivation

One of the most significant contributions of fitness and health trackers is their psychological influence. The act of tracking itself often increases awareness and fosters motivation, a phenomenon known as the Hawthorne effect. Gamification features such as badges, streaks, and virtual competitions further enhance motivation by making health goals feel engaging and rewarding. Social sharing options allow users to connect with friends, fostering accountability and community support. However, the psychological effects can also be complex, as some individuals may feel pressure or anxiety when they fail to meet goals. Striking a balance between motivation and flexibility is essential to ensure that tracking supports long-term wellness rather than becoming a source of stress.

Fitness Trackers in Athletic Performance

For athletes and fitness enthusiasts, trackers offer valuable insights that go beyond general wellness. Real-time heart rate monitoring allows athletes to train within specific zones that maximize endurance, fat burning, or cardiovascular improvement. GPS tracking provides accurate data on distance, pace, and elevation, which is particularly beneficial for runners, cyclists, and hikers. Recovery features, such as heart rate variability monitoring, help athletes avoid overtraining and optimize rest periods. By analyzing this data, athletes can refine their training programs, identify weaknesses, and prevent injuries. Fitness trackers thus serve not only as motivational tools but also as performance-enhancing devices that integrate science into athletic routines.

Medical Applications and Preventive Health

Beyond fitness, health trackers are increasingly being used in medical contexts. Devices capable of recording electrocardiograms can detect irregularities such as atrial fibrillation, enabling early intervention. Blood oxygen sensors gained prominence during the COVID-19 pandemic, as they provided a non-invasive way to monitor respiratory health. Some trackers also include stress monitoring, using metrics such as heart rate variability to indicate levels of physiological strain. For patients managing chronic conditions like diabetes or hypertension, trackers provide continuous feedback that complements traditional medical care. Healthcare providers are beginning to recognize the value of this data, integrating it into preventive strategies and remote patient monitoring programs. This trend reflects a broader movement toward personalized medicine, where wearable technology plays a central role in disease prevention and management.

Limitations and Challenges

While fitness and health trackers offer many benefits, they are not without limitations. Accuracy remains a concern, as devices may vary in their ability to precisely measure steps, calories, or sleep stages. Overreliance on trackers can sometimes lead to obsessive behaviors, where individuals prioritize metrics over overall well-being. Privacy and data security are also major concerns, as these devices collect sensitive health information that could be misused if not properly protected. Moreover, not all users sustain long-term engagement, with many abandoning their trackers after initial enthusiasm fades. Addressing these challenges requires both technological improvements and a balanced approach to health tracking that emphasizes sustainability, accuracy, and data protection.

2.3.2 Continuous Glucose Monitoring (CGM)

A continuous glucose monitor (CGM) is a device for monitoring blood glucose continuously instead of monitoring periodically by drawing a drop of blood from a finger. This is known as continuous glucose monitoring. CGMs are used by people who treat their diabetes with insulin, for example people with type 1 diabetes, type 2 diabetes, or other types of diabetes, such as gestational diabetes.

A continuous glucose monitor has three parts:

- A small electrode that is placed under the skin.
- A transmitter that sends readings from the electrode to a receiver at regular intervals (every 1 to 15 minutes).
- A separate receiver that shows the glucose level on a display.

Approved CGMs use an enzymatic technology which reacts with glucose molecules in the body's interstitial fluid to generate an electric current that is proportional to glucose concentration. Data about glucose concentration is then relayed from a transmitter attached to the sensor, to a receiver that displays the data to the user. Some CGM devices must be calibrated periodically with traditional blood glucose measurements, but others do not require calibration by the user.

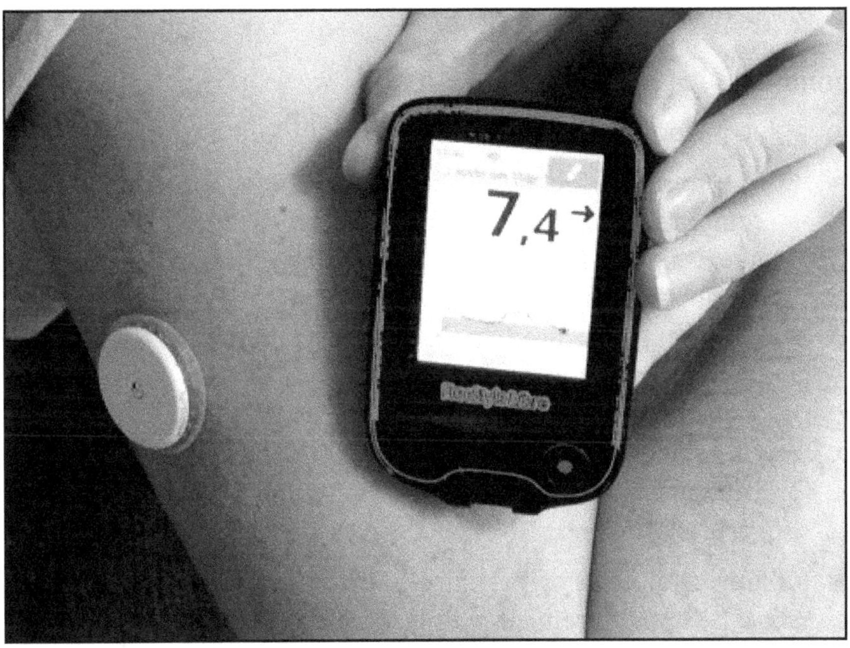

Fig: The Sensor and Transmitter are Fixed to the Upper Arm. The Receiver Shows the Blood Glucose Level and a Graph of Recent Blood Glucose Levels

Benefits

Continuous glucose monitoring is gaining in popularity for a variety of reasons.

- Traditional fingerstick testing measures blood glucose levels at only a single point in time. CGM enables users to see blood glucose levels continually, as well as trends in blood glucose levels over time.

- CGM is more convenient and less painful than traditional fingerstick testing.
- Some studies have demonstrated that CGM users spend less time in hypoglycemia or with lower glycated hemoglobin, both of which are favorable outcomes.
- Continuous glucose monitors appear to lower hemoglobin A1c levels more than just monitoring with capillary blood testing, particularly when CGM is used by people with poorly controlled diabetes and combined with the use of an integrated insulin pump. However, a Cochrane systematic review found limited and conflicting evidence of the effectiveness of continuous glucose monitoring systems in children, adults, and users with poorly controlled diabetes.
- A meta-analysis of 25 randomised controlled trials found that CGM-based feedback to support behavior change modestly reduces glycated hemoglobin levels and increases time-in-range in individuals with and without diabetes.

Limitations

Continuous glucose monitoring has some important limitations:

- CGM systems are not sufficiently accurate for detecting hypoglycemia, a common side-effect of diabetes treatment. This is especially problematic because some devices have alarm functions to warn users about a hypoglycemic condition, and people might rely on those alarms. However, the Cochrane systematic review showed that the use of continuous glucose monitors did not increase the risk of hypoglycemia or ketoacidosis. Some manufacturers warn users about relying solely on CGM measurements. The National Institute for Health and Care Excellence recommends validating hypoglycaemic values with fingerprick testing.
- Compression lows or pressure-induced sensitivity attenuations (PISA) are false hypoglycemic readings resulting from pressure applied at the CGM site. These can be caused by the user sleeping or sitting on the sensor, and may result in incorrect treatment.
- Another limitation of CGM is that glucose levels are taken from the body's interstitial fluid rather than from the blood. Because it takes time for glucose to travel from the bloodstream into the interstitial fluid, there is an inherent lag between actual blood glucose level and the level measured by the CGM. This lag time varies by both user and device, but it is usually 5 to 20 minutes.

Flash Glucose Monitoring

The original Freestyle Libre monitor introduced by Abbott Diabetes Care in 2015 was described as doing "flash glucose monitoring," with a disposable 14-day sensor probe under the skin (as with other CGM sensors), but factory-calibrated without requiring calibration against a fingerstick glucose test.

The sensor measures the glucose level of interstitial fluids (as a proxy for blood sugar levels) continuously; up to eight hours of these readings, averaged over each 15-minute period, are stored in the sensor unit, unlike most other CGM systems, which use a wireless link (typically Bluetooth) to an external device for each reading. Data stored in the sensor are transmitted on demand to a "reader" held within a centimeter or two of the sensor unit, employing near-field communication (NFC) technology. As only eight hours worth of data can be stored, downloads must not be spaced more than eight hours apart.

Differences in US insurance coverage favoring "flash glucose monitoring" over "continuous glucose monitoring" were an advantage to early adoption of Abbott's less expensive system. In the UK, flash glucose monitors and sensors are available to many patients without charge on the National Health Service (NHS).

The later Freestyle Libre 2 version of Abbott's device uses different, incompatible, sensors. It can be programmed to transmit a low blood sugar (hypoglycemia) or high sugar warning via Bluetooth to a nearby device and, as of 2023, transmits glucose readings via Bluetooth on a 60-second basis effectively making a CGM and not a flash glucose monitor. The following Freestyle Libre 3 is smaller, and transmits its readings via Bluetooth, as other meters do; it is not described as flash monitoring.

Device Characteristics

- Continuous versus flash monitoring: Dexcom, Eversense, and Libre 2 and 3 use continuous monitoring where information on the glucose levels are continuously updated. Continuous monitoring allows to set automatic alarms that are triggered when the glucose level goes out of pre-configured thresholds. In contrast, with flash monitoring such as the Freestyle Libre1, the glucose level is read automatically by the sensor; however, data is only transmitted to the user on user request. The glucose information stored on the sensor contains all the data since the previous read (up to 8 hours). FreeStyle Libre 2 allows configuration of alarms when glucose reaches a pre-determined level.
- Implantable sensors: Since the electronics and battery require a relatively large package, most CGM sensors are worn over the skin with the actual sensing probe penetrating the skin. However the Eversense

sensor is an actual implant, and receives its power wirelessly from a so-called transmitter worn above the skin. The "transmitter" receives data from the sensor every 5 minutes and forwards that data to a nearby device wirelessly. However unlike the Freestyle Libre, the implanted device is too small to have its own battery and memory, so that no glucose readings are generated during periods in which the transmitter is not being worn. The transmitter must be removed at least once a day for recharging (10 minutes) and replacement of the adhesive.

Closed-Loop System

The CGM is a key element in the development of a "closed-loop" system for the treatment of type I diabetes. A closed-loop system monitors blood glucose by CGM and sends data to an insulin pump for calculated delivery of insulin without user intervention. A number of insulin pumps currently offer an "auto mode" however this is not yet a fully closed loop system. There are several implementations, including the artificial pancreas system and the open source OpenAPS.

Emerging CGM Technologies

The continuous glucose monitoring space remains subject to extensive research and development in building lower cost, more accurate and more easy-to-use sensing solutions, some of which aim to be noninvasive. A noninvasive CGM has been defined as a medical device that can measure glucose levels in the body without puncturing the skin, drawing blood or causing any pain.

As of August 2023, besides Dexcom and Abbott Diabetes, no other manufacturer has attained a significant market share worldwide. There have been regulatory approvals of noninvasive sensing systems in Europe, though market adoption has been low, not affecting the Abbott-Dexcom dominance.

Emerging Invasive CGM Technologies

Multiple invasive CGM solutions have been under development since the early 2000s. Senseonics has commercialized its 180-day Eversense XL sensing systems in both the U.S. and European markets. In June 2023, it announced what it deemed favorable safety and accuracy data for its 365-day sensor, suggesting it may be commercializable in the future. In June 2024, Senseonics announced intent to launch sales of the 365-day sensor in 2025. A solution built by U.S. firm GlySens, aimed to remove the need for an external reader by creating a sensor that could be implanted under the skin, that directly transmitted glucose values to an external app. As of August 2023, this undertaking has stalled and the system has not been approved anywhere and the company is defunct. Another invasive CGM technology under development by Profusa Inc, based in Emeryville, California, builds on sensing research projects previously undertaken by the company under DARPA grants. This technology is composed of a hydrogel microsensor that is placed under the skin subcutaneously in a non-surgical procedure. In a 2020 literature review several biomedical engineers supported Profusa's claims that the non-surgical insertion procedure differentiates it favorably from Senseonics' Eversense system, as the latter requires a surgical procedure to insert and remove the sensor. The Profusa sensor allegedly also does not need to be removed because it overcomes the foreign body response. A reader is placed on the skin on top of where the sensor is, with the sensor transmitting a light signal to it. The sensor is claimed to last for three to six months. The is information then passed on to a smartphone where it can be tracked through an app. As of August 2023, this sensor has not attained regulatory approval in any jurisdiction, though a similar Profusa system measuring oxygen levels under the skin, has CE certification in Europe. Profusa has filed to go public via SPAC transaction.

A similar approach was under development by another California-based company called Metronom Health. This company has not released news releases, nor has any news covered any progress in terms of its research and development.

Yet another invasive approach is being developed by Belgium-based Indigo Diabetes. Indigo states that it is developing a CGM called a "continuous multi-metabolite monitoring system (CMM)". It is designed to provide people living with diabetes access to information on their glucose and other metabolite levels at any given time. It has yet to attain regulatory approval. The company completed a clinical trial in April 2024.

Emerging Noninvasive CGM Technologies

The ease of use many CGM users expect would be provided by a safe and accurate noninvasive device has led to significant innovation and research. Noninvasive approaches can be divided into interstitial fluid-based, radio frequency-based or breath-based. Interstitial fluid-analyzing sensors either use a device to analyze fluid on the skin or under the skin by sending infrared lasers to detect glucose levels in fluid. Radio frequency devices go through the skin and may derive glucose level information from blood directly. Apple has reportedly been working on a noninvasive CGM that it seeks to integrate into its Apple Watch. In March 2023 it was reported to have established proof-of-concept of a noninvasive CGM. Another company working on noninvasive CGM is Masimo, which sued Apple for patent infringement in this area in 2020. Masimo has also filed new patents through its subsidiary Cercacor (pending as of September 2023) covering a joint continuous glucose monitoring and pump-closed loop delivery system. Samsung announced that it would be incorporating glucose monitoring with its smartwatch with a targeted release year of 2025. As of October 2023 the last update was in December 2022. It is not clear whether the watch will integrate readings from an external CGM such as Dexcom's or Abbott's, or work standalone. The company in 2020 published literature regarding a non-invasive method it had developed with MIT scientists to engage in continuous glucose monitoring using spectroscopy. The company has filed patents related to this technology.

SugarBeat, built by Nemaura Medical, is a wireless non-invasive blood glucose monitoring system using a disposable skin patch. The patch connects to a rechargeable transmitter which detects blood sugar and transfers the data to a mobile app every five minutes. The patch can be used for 24 hours. Electronic currents are used to draw interstitial fluid to the surface to analyse the glucose level. SugarBeat has achieved regulatory approval in Saudi Arabia and Europe, though market penetration rates remain very low. The company declared US$503,906 in revenue for the fiscal year ending March 2022, which compares to Dexcom's more than $3 billion. As of August 2023 it had submitted a US FDA premarket approval application for sugarBEAT.

Another noninvasive system is built by US company Movano Health. It uses a small ring placed on the arm. Movano said in 2021 that it was building the smallest ever custom radio frequency (RF)-enabled sensor designed for simultaneous blood pressure and glucose monitoring. Movano is listed as MOVE on NASDAQ. By August 2023 Movano had shifted to building sensor rings for other parameters, such as heart rate, blood oxygen levels, respiration rate, skin temperature variability, and menstrual symptom tracking.

DiaMonTech AG is a Berlin, Germany-based privately held company developing the D-Pocket, a non-CGM glucose sensor that uses infrared laser technology to scan the tissue fluid in the skin and detect glucose molecules. Short pulses of infrared light are sent to the skin, which are absorbed by the glucose molecules. This generates heat waves that are detected using its patented IRE-PTD method. The company claims a high selectivity of its method, results of a first study have been published in the Journal of Diabetes Science and Technology. In this study, a Median Absolute Relative Difference of 11.3% is claimed. DiaMonTech has announced that its envisioned follow-up product D-Sensor, will feature continuous measurements, making it a CGM though no release date has been given.

The BioXensor developed by British company BioRX uses patented radio frequency technology, alongside a multiple sensor (also capturing blood oxygen levels, ECG, respiration rate, heart rate and body temperature) approach. The company claims this enables the measurement of blood glucose levels every minute reliably, accurately, and non-invasively. BioXensor had not received regulatory approval as of June 2023. Haifa, Israel-based company HAGAR completed a study of its GWave non-invasive CGM, reporting high accuracy. This sensor uses radiofrequency waves to measure glucose levels in the blood. The device had not received regulatory approval anywhere as of August 2023. One of the criticisms of radiofrequency technology as a way of measuring glucose is that studies in 2019 found that glucose can only be detected in the far infrared (nanometer wavelengths), rather than radiofrequencies even in the centimeter and millimeter wavelength range, putting into question the viability of radio frequencies for measuring glucose.

Glucomodicum is based in Helsinki, Finland. Their attempted solution uses interstitial fluid to non-invasively measure glucose levels continuously. It does not have regulatory approval. KnowLabs is a Seattle, U.S-based company building a CGM called the Bio-RFID sensor, which works by sending radio waves through the skin to measure molecular signatures in the blood, which Know Labs' machine learning algorithms use to compute the user's blood sugar levels. The company reported that it had built a prototype, but had not attained regulatory approval as of August 2023. Liom (formerly named Spiden) is a Swiss startup building a multi-biomarker and drug level monitoring noninvasive smartwatch wearable with continuous glucose monitoring capability as its first application. It has so far not attained regulatory approval as of October 2023. In January of 2024, Liom (then called Spiden) declared it had developed a prototype, with a claimed MARD (Mean Absolute Relative Difference) value to a reference glucose measurement of approximately 9%. Occuity, a Reading, UK-based startup is taking a different approach to noninvasive glucose monitoring, by using the eye. The company is developing the Occuity Indigo, which will measure the change in refractive index of the eye to determine the concentration of glucose in the blood.

2.3.3 HRV and Sleep Tracking Devices

Heart Rate Variability

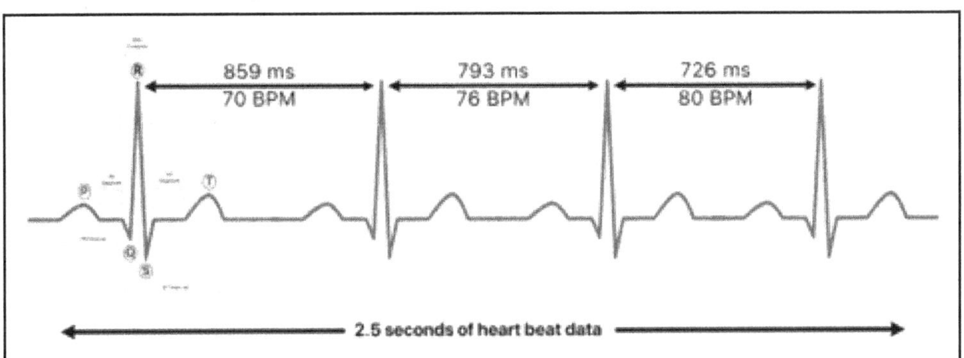

Fig: Heart Rate Variability Visualized with R-R Interval Changes

Heart rate variability (HRV) is the physiological phenomenon of variation in the time interval between heartbeats. It is measured by the variation in the beat-to-beat interval. Other terms used include "cycle length variability", "R–R variability" (where R is a point corresponding to the peak of the QRS complex of the ECG wave;

and R–R is the interval between successive Rs), and "heart period variability". Measurement of the RR interval (often termed normal-to-normal or NN interval when additional filtering is used) is used to derive heart rate variability. Methods used to detect beats include ECG, blood pressure, ballistocardiograms, and the pulse wave signal derived from a photoplethysmograph (PPG). ECG is considered the gold standard for HRV measurement because it provides a direct reflection of cardiac electric activity.

Variability

Variability (or Variation) in the beat-to-beat interval is a physiological phenomenon. Power spectral analysis of the beat-to-beat variations of heart rate or the heart period (R–R interval) partitions the total variance (the "power") of a continuous series of beats into its frequency components, typically identifying two or three main peaks: Very Low Frequency (VLF) <0.04 Hz, Low Frequency (LF), 0.04–0.15 Hz, and High Frequency (HF) 0.15–0.4 Hz. The HF peak is widely believed to reflect cardiac parasympathetic nerve (PSNS) activity while the LF, although more complex, is often assumed to have a dominant sympathetic (SNS) component. Though LF was previously thought to reflect primarily SNS activity, it is now widely accepted that it reflects a mixture of both the SNS and PSNS.

Decreased PSNS activity or increased SNS activity will result in reduced HRV. High frequency (HF) activity (0.15 to 0.40 Hz), especially, has been linked to PSNS activity. Activity in this range is associated with the respiratory sinus arrhythmia (RSA), a vagally mediated modulation of heart rate (which increases during inspiration and decreases during expiration). Less is known about the physiological inputs of the low frequency (LF) activity (0.04 to 0.15 Hz). The SA node receives several different inputs and the instantaneous heart rate or RR interval and its variation are the results of such inputs. The main inputs are the sympathetic and the parasympathetic nervous system (PSNS) and humoral factors. Respiration gives rise to waves in heart rate mediated primarily via the PSNS, and it is thought that the lag in the baroreceptor feedback loop may give rise to 10 second waves in heart rate (associated with Mayer waves of blood pressure), but this remains controversial. Factors that affect the input are:

- Baroreflex.
- Thermoregulation.
- Hormones.
- Sleep–wake cycle.
- Meals.
- Physical Activity.
- Stress.

Clinical Significance

Reduced HRV has been shown to be a predictor of mortality after myocardial infarction although others have shown that the information in HRV relevant to acute myocardial infarction survival is fully contained in the mean heart rate. A range of other outcomes and conditions may also be associated with modified (usually lower) HRV, including congestive heart failure, diabetic neuropathy, post–cardiac-transplant depression, susceptibility to SIDS and poor survival in premature babies, as well as fatigue severity in chronic fatigue syndrome. On the other hand, for patients having high blood pressure (hypertension), higher HRV is a risk factor for atrial fibrillation.

Psychological and Social Aspects

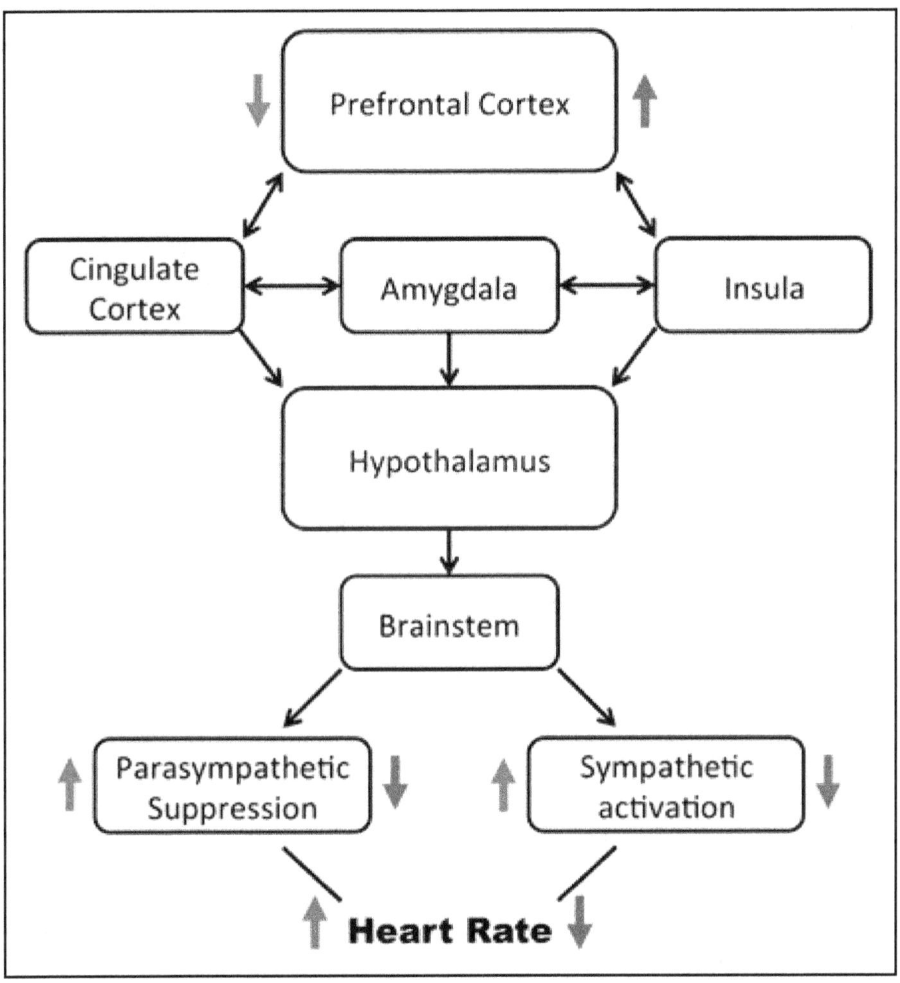

Fig: A Simplified Representation of the Neurovisceral Integration Model

There is interest in HRV in the field of psychophysiology. For example, HRV is related to emotional arousal. High-frequency (HF) activity has been found to decrease under conditions of acute time pressure and emotional strain and elevated anxiety state, presumably related to focused attention and motor inhibition. HRV has been shown to be reduced in individuals reporting to worry more. In individuals with post-traumatic stress disorder (PTSD), HRV and its HF component is reduced whilst the low-frequency (LF) component is elevated. Furthermore, PTSD patients demonstrated no LF or HF reactivity to recalling a traumatic event. Statistical quantitative differences have also been found among healthy, depressed, and psychotic people.

The neurovisceral integration is a model of HRV that views the central autonomic network as the decision maker of cognitive, behavioral and physiological regulation as they pertain to a continuum of emotion. The neurovisceral integration model describes how the prefrontal cortex regulates activity in limbic structures which act to suppress parasympathetic nervous system (PSNS) activity and activate sympathetic nervous system (SNS) circuits. Variation in the output of these two branches of the autonomic system produces HRV and activity in the prefrontal cortex can hence modulate HRV.

HRV is reported to be an index of the influence of both the parasympathetic and the sympathetic nervous system. For example, high HRV is shown to reflect proper emotion regulation, decision-making, and attention, and low HRV reflects the opposite. The parasympathetic nervous system works to decrease the heart rate, while the SNS works to increase the heart rate. For example, someone with high HRV may reflect increased parasympathetic activity, and someone with low HRV may reflect increased sympathetic activity.

Emotions stem from the time and impact of a situation on a person. The ability to regulate emotions is essential for social environments and well-being. HRV has provided a window to the physiological components associated with emotional regulation. HRV has been shown to reflect emotional regulation at two different levels, while resting and while completing a task. Research suggests that a person with higher HRV while resting can provide more appropriate emotional responses compared to those that have low HRV at rest.

Empirical research found that HRV can reflect better emotional regulation by those with higher resting HRV, particularly with negative emotions. However, HRV is elevated by negative news in persons who react more strongly to negative news than to positive news. When completing a task, HRV is subject to change, especially when people need to regulate their emotions. Most importantly, individual differences are related to the ability to regulate emotions. Previous research has suggested that a large part of the attention regulation is due to the default inhibitory properties of the prefrontal cortex. Top-down processes from the prefrontal cortex provide parasympathetic influences, and if for some reason, those influences are active, attention can suffer. Researchers have suggested that HRV can index attention. It was found that groups with high anxiety and low HRV have poor attention. In line with this research, it has also been suggested that increased attention has been linked to high HRV and increased vagus nerve activity.

The vagus nerve activity reflects the physiological modulation of the parasympathetic and sympathetic nervous system. The activity behind the prefrontal cortex and the parasympathetic and sympathetic nervous system can influence heart activity. However, people are not all affected the same. A systematic review of HRV and cognitive function suggested that resting HRV can predict individual differences in attentional performance. Furthermore, HRV has been able to index the role of attention and performance, supporting high HRV as a biomarker of increased attention and performance.

Decision-making skills are found to be indexed by HRV in several studies. Previous research has suggested that both emotion and attention are linked to decision making; for example, poor decision making is linked to the inability to regulate or control emotions and attention and vice versa. Decision making is negatively affected by lower HRV and positively affected by higher levels of HRV.

Most importantly, resting-state HRV was found to be a significant predictor of cognitive functions such as decision making. HRV, accompanied by a psychological state such as anxiety, has been found to lead to poor decisions. For example, a group of researchers found that low HRV was an index of higher uncertainty leading to poor decision-making skills, especially those with higher levels of anxiety. HRV was also used to assess decision-making skills in a high-risk game and was found to be an index higher sympathetic activation (lower HRV) when making decisions involving risk. HRV can index psychological concepts, such as the ones outlined above, to assess the demand for the situations that people experience.

The polyvagal theory is another way to describe the pathways in the autonomic nervous system that mediate HRV. The polyvagal theory highlights three main ordinal processes, inactive response to an environmental threat, the active response to an environmental threat, and the fluctuation between the connect and disconnect to an environmental threat. This theory, like others, decomposes heart rate variability based on frequency domain characteristics. However, it places more emphasis on respiratory sinus arrhythmia and its transmission by a hypothesized neural pathway distinct from other components of HRV. There is anatomic and physiological evidence for a polyvagal control of the heart.

Phenomena

There are two primary fluctuations:

- Respiratory arrhythmia (or respiratory sinus arrhythmia) is directly caused by the central respiratory rhythm, which tracks the respiratory rate across a range of frequencies and is a major cause of heart rate variability in humans. The respiratory rhythm contributes to sinus arrhythmia in normal unanesthetized subjects during mechanical hyperventilation with positive pressure.

- Low-frequency oscillations are associated with Mayer waves (Traube–Hering–Mayer waves) of blood pressure and is usually at a frequency of 0.1 Hz, or a 10-second period.

Pattern Formation in Heart Rate

The regulation of heart rate, encompassing both acceleration and deceleration of the heartbeat, is highly specific. Distinct patterns in heart rate can be observed during everyday movements, such as kneeling down or standing up. These patterns reflect highly sensitive physiological regulatory mechanisms that enable a healthy heart to adapt to various life influences.

Artifact

Errors in the location of the instantaneous heart beat will result in errors in the calculation of the HRV. HRV is highly sensitive to artifact and errors in as low as even 2% of the data will result in unwanted biases in HRV calculations. To ensure accurate results therefore it is critical to manage artifact and RR errors appropriately prior to performing any HRV analyses.

Robust management of artifacts, including RWave identification, interpolation and exclusion requires a high degree of care and precision. This can be very time-consuming in large studies with data recorded over long durations. Software packages are able to assist users with a variety of robust and tested artifact management tools. These software programs also include some automated capability but it is important that a human review any automated artifact management and edit accordingly.

Analysis

The most widely used methods can be grouped under time-domain and frequency-domain. A joint European and American task-force described standards in HRV measurements in 1996. Other methods have been proposed, such as non-linear methods.

Time-Domain Methods

Time-domain methods are based on the beat-to-beat or NN intervals (typically determined from RR intervals), which are analysed to give variables such as:

- SDNN (standard deviation of NN intervals). Often calculated over a 24-hour period. SDNN reflects all the cyclic components responsible for variability in the period of recording, therefore it represents total variability.

- SDANN (standard deviation of the average NN intervals) calculated over short periods, usually 5 minutes. SDANN is therefore a measure of changes in heart rate due to cycles longer than 5 minutes.
- RMSSD (root mean square of successive differences), the square root of the mean of the squares of the successive differences between adjacent NNs.
- SDSD (standard deviation of successive differences), the standard deviation of the successive differences between adjacent NNs.
- NN50, the number of pairs of successive NNs that differ by more than 50 ms.
- pNN50, the proportion of NN50 divided by total number of NNs.
- NN20, the number of pairs of successive NNs that differ by more than 20 ms.
- pNN20, the proportion of NN20 divided by total number of NNs.
- EBC (estimated breath cycle), the range (max-min) within a moving window of a given time duration within the study period. The windows can move in a self-overlapping way or be strictly distinct (sequential) windows. EBC is often provided in data acquisition scenarios where HRV feedback in real time is a primary goal. EBC derived from PPG over 10-second and 16-second sequential and overlapping windows has been shown to correlate highly with SDNN.

Geometric Methods

The series of NN intervals also can be converted into a geometric pattern such as: Geometric Measures HRV triangular index: integral of density distribution / maximum of density distribution maximum HRV triangular index = Number of all NN intervals / maximum number. Dependent on the length of the bin -> quote the bin size+ relative insensitive to the analytic quality of the series of NN intervals – need of reasonable number of NN intervals to generate the geometric pattern (in practice 20 min to 24 h) – not appropriate to assess short-term changes in HRV

- The sample density distribution of NN interval durations;
- Sample density distribution of differences between adjacent NN intervals; and
- A scatterplot of each NN (or RR) interval with the immediately preceding NN (or RR) interval – also called "Poincare plot" or (apparently in error)

a "Lorenz plot"; and so forth. A simple formula is then used that judges the variability on the basis of the geometric and/or graphics properties of the resulting pattern.

Frequency-Domain Methods

Frequency domain methods assign bands of frequency and then count the number of NN intervals that match each band. The bands are typically high frequency (HF) from 0.15 to 0.4 Hz, low frequency (LF) from 0.04 to 0.15 Hz, and the very low frequency (VLF) from 0.0033 to 0.04 Hz. HF power reflects stimulation by the parasympathetic nervous system (PNS), whereas LF power reflects stimulation by both the sympathetic nervous system (SNS) and the PNS.

VLF power is associated with thermoregulation, the renin–angiotensin system. and peripheral vasomotor activity. Several methods of analysis are available. Power spectral density (PSD), using parametric or nonparametric methods, provides basic information on the power distribution across frequencies. One of the most commonly used PSD methods is the discrete Fourier transform. Methods for the calculation of PSD may be generally classified as nonparametric and parametric. In most instances, both methods provide comparable results. The advantages of the nonparametric methods are:

- The simplicity of the algorithm used (fast Fourier transform [FFT] in most of the cases) and
- The high processing speed.

The advantages of parametric methods are:

- Smoother spectral components that can be distinguished independent of preselected frequency bands,
- Easy postprocessing of the spectrum with an automatic calculation of low- and high-frequency power components with an easy identification of the central frequency of each component, and
- An accurate estimation of PSD even on a small number of samples on which the signal is supposed to maintain stationarity.

The basic disadvantage of parametric methods is the need of verification of the suitability of the chosen model and of its complexity (that is, the order of the model). In addition to classical FFT-based methods used for the calculation of frequency parameters, a more appropriate PSD estimation method is the Lomb–Scargle periodogram. Analysis has shown that the LS periodogram can produce a more accurate estimate of the PSD than FFT methods for typical RR data. Since the RR data is an unevenly sampled data, another advantage of the LS method is that in contrast to FFT-based methods it is able to be used without the need to resample and detrend the RR data. Alternatively, to avoid artefacts that are created when calculating the power of a signal that includes a single high-intensity peak (for example caused by an arrhythmic heart beat), the concept of the 'instantaneous Amplitude' has been introduced, which is based on the Hilbert transform of the RR data.

A newly used HRV index, which depends on the wavelet entropy measures, is an alternative choice. The wavelet entropy measures are calculated using a three-step procedure defined in the literature. First, the wavelet packet algorithm is implemented using the Daubechies 4 (DB4) function as the mother wavelet with a scale of 7. Once the wavelet coefficients are obtained, the energy for each coefficient are calculated as described in the literature. After calculating the normalized values of wavelet energies, which represent the relative wavelet energy (or the probability distribution), the wavelet entropies are obtained using the definition of entropy given by Shannon.

Non-Linear Methods

Given the complexity of the mechanisms regulating heart rate, it is reasonable to assume that applying HRV analysis based on methods of non-linear dynamics will yield valuable information. Although chaotic behavior has been assumed, more rigorous testing has shown that heart rate variability cannot be described as a low dimensional chaotic process. However, application of chaotic globals to HRV has been shown to predict diabetes status. The most commonly used non-linear method of analysing heart rate variability is the Poincaré plot. Each data point represents a pair of successive beats, the x-axis is the current RR interval, while the y-axis is the previous RR interval. HRV is quantified by fitting mathematically defined geometric shapes to the data. Other methods used are the correlation dimension, symbolic dynamics, nonlinear predictability, pointwise correlation dimension, approximate entropy, sample entropy, multiscale entropy analysis, sample asymmetry and memory length (based on inverse statistical analysis). It is also possible to represent long range correlations geometrically.

Long Term Correlations

Sequences of RR intervals have been found to have long-term correlations. However, one flaw with these analyses is their lack of goodness-of-fit statistics, i.e. values are derived that may or may not have adequate statistical rigor. Different types of correlations have been found during different sleep stages.

Heart Rate Dependence of HRV Parameters

A basic problem is that all the parameters used to characterize HRV strongly depend on heart rate, but many articles have not adjusted properly or at all for HR differences when comparing HRV in multiple circumstances.

However, the exact HRV(HR) relationship is still a matter of debate. For time domain parameters (RMSSD, SDNN, etc.) the results imply that, if there exists a universal function, it should be either exponential or hyperbolic in nature. The evaluation procedures used to determine HRV(HR) function have not allowed to decide between these options, so far.

A new evaluation method has recently allowed to determine a HRV(HR) function with unprecedented precision: it can be described by two descending exponential components for healthy individuals, in general.

Duration and Circumstances of ECG Recording

Time domain methods are preferred to frequency domain methods when short-term recordings are investigated. This is due to the fact that the recording should be at least 10 times the wavelength of the lowest frequency bound of interest. Thus, recording of approximately 1 minute is needed to assess the HF components of HRV (i.e., a lowest bound of 0.15 Hz is a cycle of 6.6 seconds and so 10 cycles require ~60 seconds), while more than 4 minutes are needed to address the LF component (with a lower bound of 0.04 Hz).

Although time domain methods, especially the SDNN and RMSSD methods, can be used to investigate recordings of long durations, a substantial part of the long-term variability is day–night differences. Thus, long-term recordings analyzed by time domain methods should contain at least 18 hours of analyzable ECG data that include the whole night.

Physiological Correlates of HRV Components

Autonomic Influences of Heart Rate

Although cardiac automaticity is intrinsic to various pacemaker tissues, heart rate and rhythm are largely under the control of the autonomic nervous system. The parasympathetic influence on heart rate is mediated via release of acetylcholine by the vagus nerve. Muscarinic acetylcholine receptors respond to this release mostly by an increase in cell membrane K+ conductance. Acetylcholine also inhibits the hyperpolarization-activated "pacemaker" current. The "Ik decay" hypothesis proposes that pacemaker depolarization results from slow deactivation of the delayed rectifier current, Ik, which, due to a time-independent background inward current, causes diastolic depolarization. Conversely, the "If activation" hypothesis suggests that after action potential termination, If provides a slowly activating inward current predominating over decaying Ik, thus initiating slow diastolic depolarization. The sympathetic influence on heart rate is mediated by release of epinephrine and norepinephrine. Activation of β-adrenergic receptors results in cAMP-mediated phosphorylation of membrane proteins and increases in ICaL and in If the result is an acceleration of the slow diastolic depolarization.

Under resting conditions, vagal tone prevails and variations in heart period are largely dependent on vagal modulation. The vagal and sympathetic activity constantly interact. Because the sinus node is rich in acetylcholinesterase, the effect of any vagal impulse is brief because the acetylcholine is rapidly hydrolyzed. Parasympathetic influences exceed sympathetic effects probably through two independent mechanisms: a cholinergically induced reduction of norepinephrine released in response to sympathetic activity, and a cholinergic attenuation of the response to an adrenergic stimulus.

Components

The RR interval variations present during resting conditions represent beat-by-beat variations in cardiac autonomic inputs. However, efferent vagal (parasympathetic) activity is a major contributor to the HF component, as seen in clinical and experimental observations of autonomic maneuvers such as electrical vagal stimulation, muscarinic receptor blockade, and vagotomy. More problematic is the interpretation of the LF component, which was considered by some as a marker of sympathetic modulation (especially when expressed in normalized units) but is now known to include both sympathetic and vagal influences. For example, during sympathetic activation the resulting tachycardia is usually accompanied by a marked reduction in total power, whereas the reverse occurs during vagal activation. Thus the spectral components change in the same direction and do not indicate that LF faithfully reflects sympathetic effects.

HRV measures fluctuations in autonomic inputs to the heart rather than the mean level of autonomic inputs. Thus, both withdrawal and saturatingly high levels of autonomic input to the heart can lead to diminished HRV.

Changes Related to Biologic States and Pathologies

A reduction of HRV has been reported in several cardiovascular and noncardiovascular diseases.

- **Myocardial Infarction:** Depressed HRV after MI may reflect a decrease in vagal activity directed to the heart. HRV in patients surviving an acute MI reveal a reduction in total and in the individual power of spectral components. The presence of an alteration in neural control is also reflected in a blunting of day-night variations of RR interval. In post-MI patients with a very depressed HRV, most of the residual energy is distributed in the VLF frequency range below 0.03 Hz, with only a small respiration-related variations.
- **Diabetic Neuropathy:** In neuropathy associated with diabetes mellitus characterized by alteration in small nerve fibers, a reduction in time domain parameters of HRV seems not only to carry negative prognostic

value but also to precede the clinical expression of autonomic neuropathy. In diabetic patients without evidence of autonomic neuropathy, reduction of the absolute power of LF and HF during controlled conditions was also reported. Similarly, diabetic patients can be differentiated from normal controls on the basis of reduction in HRV.

- **Heart Transplants:** A very reduced HRV with no definite spectral components has been reported in patients with a recent heart transplant. The appearance of discrete spectral components in a few patients is considered to reflect cardiac reinnervation. This reinnervation may occur as early as 1 to 2 years after transplantation and is assumed to be of sympathetic origin. In addition, a correlation between respiratory rate and the HF component of HRV observed in some transplanted patients also indicates that a nonneural mechanism may generate a respiration-related rhythmic oscillation.
- **Myocardial Dysfunction:** A reduced HRV has been observed consistently in patients with cardiac failure. In this condition characterized by signs of sympathetic activation such as faster heart rates and high levels of circulating catecholamines, a relation between changes in HRV and the extent of left ventricular dysfunction was reported. In fact, whereas the reduction in time domain measures of HRV seemed to parallel the severity of the disease, the relationship between spectral components and indices of ventricular dysfunction appears to be more complex. In particular, in most patients with a very advanced phase of the disease and with a drastic reduction in HRV, an LF component could not be detected despite the clinical signs of sympathetic activation. This reflects that, as stated above, the LF may not accurately reflect cardiac sympathetic tone.
- **Liver Cirrhosis:** Liver cirrhosis is associated with decreased HRV. Decreased HRV in patients with cirrhosis has a prognostic value and predicts mortality. Loss of HRV is also associated with higher plasma pro-inflammatory cytokine levels and impaired neurocognitive function in this patient population.
- **Sepsis:** HRV is decreased in patients with sepsis. Loss of HRV has both diagnostic and prognostic value in neonates with sepsis. The pathophysiology of decreased HRV in sepsis is not well understood but there is experimental evidence to show that partial uncoupling of cardiac pacemaker cells from autonomic neural control may play a role in decreased HRV during acute systemic inflammation. (Decreased HRV is generally lower in inflammatory conditions).

- **Tetraplegia:** Patients with chronic complete high cervical spinal cord lesions have intact efferent vagal neural pathways directed to the sinus node. However, an LF component can be detected in HRV and arterial pressure variabilities of some tetraplegic patients. Thus, the LF component of HRV in those without intact sympathetic inputs to the heart represent vagal modulation.
- **Sudden Cardiac Death:** Victims of sudden cardiac death have been found to have had lower HRV than healthy individuals. HRV can be observed to be depressed prior to the development of SCD, which raises questions about whether or not altered autonomic function plays a role in the development of electrical instability. HRV is also depressed in SCD survivors, who are at high risk for subsequent episodes. HRV is markedly decreased prior to both fatal and non-fatal arrhythmias.
- **Cancer:** HRV correlates with the progression of disease and outcome of cancer patients, according to a systematic review of published studies. Patients in the early stages of cancer have a significantly higher HRV when compared to patients in the later stages of cancer, suggesting disease severity influences HRV. Different ranges of HRV can be observed between cancer types.
- **Pregnancy:** HRV alterations occur in healthy pregnancies as well as similar changes in pregnancies with gestational diabetes that include lower HRV mean values.
- **Mood and Anxiety Disorders:** Low RMSSD, thought to represent vagal tone, have been associated with major depression. Lower SDNN and elevated LF/HF were found in those with bipolar disorder, and in particular those characterized as having greater illness severity due to greater number of episodes, illness duration and whether there had been psychosis. Patients with PTSD also had lower HF, a measure of vagal tone.

Modifications by Specific Interventions

Interventions that augment HRV may be protective against cardiac mortality and sudden cardiac death. Although the rationale for changing HRV is sound, it also contains the inherent danger of leading to the unwarranted assumption that modification of HRV translates directly into cardiac protection, which may not be the case. Despite the growing consensus that increases in vagal activity can be beneficial, it is not as yet known how much vagal activity (or HRV as a marker) has to increase in order to provide adequate protection.

- **β-Adrenergic Blockade:** The data on the effect of β-blockers on HRV in post-MI patients are surprisingly scant. Despite the observation of statistically significant increases, the actual changes are very modest. In conscious post-MI dogs, β-blockers do not modify HRV. The unexpected observation that before MI, β-blockade increases HRV only in the animals destined to be at low risk for lethal arrhythmias after MI may suggest novel approaches to post-MI risk stratification.

- **Antiarrhythmic Drugs:** Data exist for several antiarrhythmic drugs. Flecainide and propafenone but not amiodarone were reported to decrease time domain measures of HRV in patients with chronic ventricular arrhythmia. In another study, propafenone reduced HRV and decreased LF much more than HF. A larger study confirmed that flecainide, also encainide and moricizine, decreased HRV in post-MI patients but found no correlation between the change in HRV and mortality during follow-up. Thus, some antiarrhythmic drugs associated with increased mortality can reduce HRV. However, it is not known whether these changes in HRV have any direct prognostic significance.
- **Scopolamine:** Low-dose muscarinic receptor blockers, such as atropine and scopolamine, may produce a paradoxical increase in vagal effects on the heart, as suggested by a decrease in heart rate. In addition, scopolamine and low dose atropine can markedly increase HRV. However, though the heart rate slowing in proportional to the (low) dose of atropine, the increase in HRV varies widely across and within individuals. This suggests that even for vagal activity to the heart, HRV may be a limited marker.
- **Thrombolysis:** The effect of thrombolysis on HRV (assessed by pNN50) was reported in 95 patients with acute MI. HRV was higher 90 minutes after thrombolysis in the patients with patency of the infarct-related artery. However, this difference was no longer evident when the entire 24 hours were analyzed.
- **Exercise-Training:** Exercise training may decrease cardiovascular mortality and sudden cardiac death. Regular exercise training is also thought to modify cardiac autonomic control. Individuals who exercise regularly have a 'training bradycardia' (i.e., low resting heart rate) and generally have higher HRV than sedentary individuals.
- **Biofeedback:** The technique called resonant breathing biofeedback teaches how to recognize and control involuntary heart rate variability. A randomized study by Sutarto et al. assessed the effect of resonant breathing biofeedback among manufacturing operators; depression, anxiety and stress significantly decreased. A first overall meta-analysis by Goessl VC et al. (24 studies, 484 individuals, 2017) indicates ''HRV

biofeedback training is associated with a large reduction in self-reported stress and anxiety", while mentioning that more well-controlled studies are needed.

- **Wind Instruments:** One study that surveyed the physiological effects of playing Native American flutes found a significant HRV increase when playing both low-pitched and high-pitched flutes.

Normal HRV Values

Even though there are no widely accepted standard values for HRV that can be used for clinical purposes, multiple studies have measured and reported normal values for various populations:

Table 1: Normal Values of Standard Measures of HRV

Time Domain Analysis				Spectral Analysis		
Variable	Units	Normal Values (mean±SD)		Variable	Units	Normal Values (mean±SD)
IBI	ms	926±90		LF	ms^2	519±291
SDNN	ms	50±16		HF	ms^2	657±777
RMSSD	ms	42±15		LF	nu	52±10
				HF	nu	40±10
				LF/HF ratio		2.8±2.6

- **Time Domain Analysis Abbreviations:** IBI is InterBeat Interval, the time period between successive heartbeats (Normal-to-Normal interbeat interval, also known as the R–R interval), measured in milliseconds (ms). SDNN is Standard Deviation of Normal-to-Normal interbeat intervals measured in milliseconds. RMSSD is Root Mean Square of Successive Differences between normal heartbeats measured in milliseconds. The conventional recording time is five minutes.

- **Spectral Analysis Abbreviations:** LF is Power of the Low Frequency range [ms squared (ms^2) or normal units (nu)] HF is Power of the High Frequency range [ms squared (ms^2) or normal units (nu)] LF/HF is the ratio of LF-to-HF power

Sleep Tracking

Sleep tracking is the process of monitoring a person's sleep, most commonly through measuring inactivity and movement. A device that tracks a person's sleep is called a sleep tracker. Sleep tracking may be beneficial in diagnosing sleep disorders. As sleep abnormalities are also symptoms of mental illness or relapsing psychotic disorders, it may also be beneficial in diagnosing mental disorders and psychotic disorders as well.

Polysomnography, the "gold standard" method for sleep tracking that requires attaching electrodes and monitors to the patient as they sleep, was developed in the late 1950s. It is considered by sleep researchers as providing the most accurate sleep data, however, it is an expensive, often uncomfortable experience for patients with findings that may be skewed due to the "first night effect". The actigraphy, a sleep-tracking device that is worn on one's wrist, was developed in the early 1970s and uses motion sensors. It is considered the "silver standard" method of sleep tracking, is comparably less expensive than a polysomnograph, and easier to incorporate into a patient's every day schedule as it looks and feels like a wrist-watch. However, it cannot track sleep-staging, is still generally expensive, and still requires a specialist to analyze the data it collects.

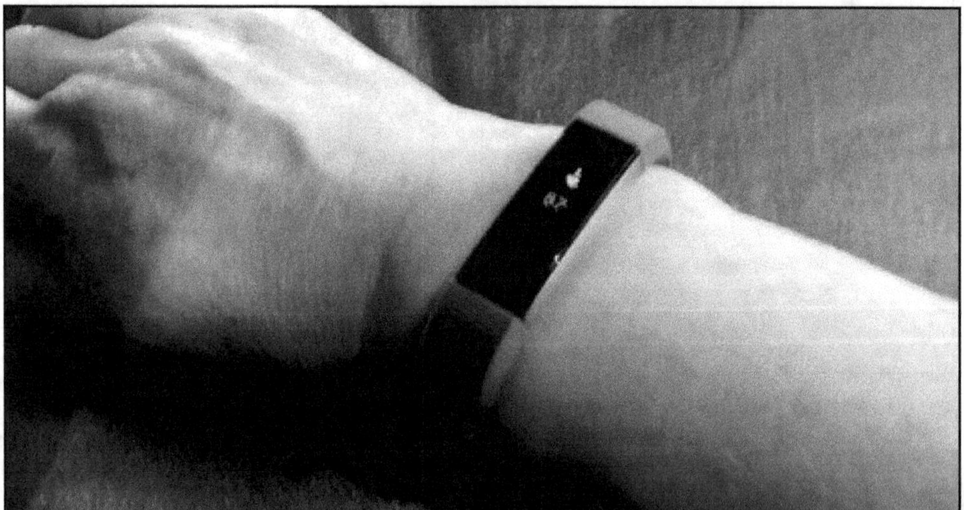

A Wearable Device Capable of Monitoring a Person's Sleep

Sleep trackers are now available to consumers in many different forms such as smartphones, smartwatches, fitness trackers, and other wearable devices. Compared to a polysomonograph or an actigraph, consumer sleep-tracking devices are already incorporated into the day-to-day lives of patients and are the most cost-effective sleep-tracking method for patients. However, consumer sleep-tracking devices as they currently are, do not provide reliable sleep data for consumers or healthcare professionals. Additionally, consumer sleep tracking devices do not share their sleep tracking methods or algorithms with the public and may unintentionally undermine the sleep recommendations of health professionals or the need to seek professional help regarding improving sleep quality.

Devices and Methods

Developed in the late 1950s, a polysomnograph, also known as a polysomnogram or a 'sleep study', is a test used to diagnose sleep disorders and is considered as the best and most reliable method to collect sleep data from individuals. A person can undergo a polysomnograph during an overnight stay in either a hospital or in a sleep center (a laboratory).

Prior to the start of the polysomnograph, electrodes are attached to the individual's scalp, chin, and outer eyelids to record signals and monitors are attached to the individual's chest to record their heart rate and track their breathing as they sleep. As the individual sleeps, they are monitored by polysomnograph technologist who will take notes on things such as changes in heart rate and breathing. Sometimes, there is also a video camera recording individual's movement as they sleep.

Developed in the early 1970s, an actigraphy device is one of the earliest devices used to track the stages of a person's sleep and identify sleep disorders. It is a non-invasive wearable device shaped like a wristwatch that tracks the movement of your body with accelerometers, small motion sensors. This device can collect data over an extended period of time, such as a few weeks or months. Actigraphy devices then uses the collected data to determine if the patient is asleep or awake and can also track the individual's other sleep behaviors such as wake time.

Sleep tracking is now possible through consumer wearable devices such as smartwatches and fitness trackers and applications on smartphones. The features that these consumer sleep-tracking devices offer can vary depending on the device, model, and version. Some sleep-tracking devices are capable of tracking the stages of a person's sleep (light sleep, deep sleep, REM sleep), the length/duration of a person's sleep, the quality of a person's sleep, and the consistency of a person's sleep.

Other features offered by sleep-tracking devices may include "sleep scores" that rank how well a person slept, "smart alarms" that wake a person up within a set period of time based on the circumstances of the person's sleep, and the ability to track the amount of light and/or the temperature in the person's bedroom. Unlike university sleep labs, which have made their sleep algorithms public for many years, the algorithms and methods of data collection used in consumer sleep-tracking devices have not been made public as they are proprietary and can also change at any point in time without notification to the users.

Utilization and Effectiveness

Sleep tracking can be used to track sleep abnormalities and the sleep quality of people and help healthcare providers diagnose their patients with sleep disorders. Sleep tracking can also be used for tracking sleep abnormalities that are symptoms of mental illness. For example, repeated sleep disturbances have been associated with increased risks of suicide, the development of mood disorders such as depression and anxiety, and relapse of psychotic disorders.

- **Polysomnography:** Polysomnographies are considered the "gold standard" for sleep data collection. However, polysomnographies do not create an environment that is conducive for sleep for most patients, especially patients who already struggle with sleep abnormalities. They are typically inaccessible to the average patient and inconvenient, as they require patients to have wires clipped to their face and monitors strapped to their body, they may need to sleep outside of their typical sleeping environment, they may be uncomfortable with knowing that as they sleep they are being monitored by the polysomongraph technologist, and the test is too expensive to obtain sleep tracking data from one patient for an extended period of time. One night of sleep tracking using a polysomnograph and a polysomongraph technologist can cost up to $2,000. There is the "one night effect" where patients will experience more difficulties with sleeping (reflected in decreased REM sleep, decreased sleep efficiency, and increased sleep latency) during their first night of sleep tracking via polysomongraphy due to the sleep tracking equipment and unusual sleeping circumstances. Additionally, the data collected during a polysomnography can also be subjected to human error.
- **Actigraphy:** Actigraphies are considered "silver standard" for sleep data collection but when compared to polysomnographies are more affordable and accessible to patients. Because of their compact design, actigraphies collect sleep tracking data over a longer period of time from patients without requiring their patients to make major changes to their day-to-day routine or their sleeping environment. The downsides of actigraphs are that they cannot track sleep staging, the storage of data collected on the device is limited, actigraphy devices are still expensive (around $1,000), and a specialist still needs to analyze the data collected from the devices to determine if there are any issues with the device's data collection.
- **Consumer Devices:** People can track their sleep through smartphones with consumer sleep-tracking apps, wearable devices, or a combination

or both consumer sleep-tracking apps and wearable devices. Consumer sleep-tracking devices such as smartphones and activity trackers were developed primarily for the use of consumers, not for clinical use or research. In contrast to polysomonographies and actigraphies, smartphones may be easier for people to utilize for sleep tracking purposes as they may already use smartphones in their day-to-day lives. Smartphones also have built-in motion-sensing accelerometers and microphone features and cloud storage. Wearable consumer-sleep tracking devices have been noted to have issues with lost or unusable data due to technical or software issues. The utilization of consumer sleep-tracking devices for clinical data collection has become more widely accepted by healthcare providers because these devices are more affordable and practical compared to polysomnographies and actigraphies. However, consumer sleep-tracking devices still have a long way to go before they produce accurate and reliable sleep-tracking information for the utilization of sleep disorder and mental illness treatment, research, and diagnoses. The motion sensors in smartphones and smartwatches are not medical-grade, smartphone data alone is not enough to capture the full picture of a patient's sleep staging, and data collected from consumer sleep-tracking devices are not reliable enough as monitoring tools as they tended to over or under-estimate data such as total sleep time. There is also a concern among health professionals that consumer sleep-tracking devices and applications may encourage consumers to self-diagnose in reaction to the results of their sleep grade, quality of sleep, or recorded hours of sleep from their consumer sleep-tracking device. Even though the average adult needs between seven and nine hours of sleep, a consumer sleep-tracking device may encourage all of their consumers to strive for eight hours of sleep in order to get a good sleep grade. This can result in consumers feeling anxious over the amount of sleep they get each night and cause them to change their sleeping behaviors in order to improve their sleep according to the consumer sleep-tracking device's algorithm. Additionally, consumers may feel content with the findings of their consumer sleep-tracking devices and forgo seeking professional help. In one study conducted by Rush University Medical College and Northwestern University's Feinberg School of Medicine, three patients who reported having unsatisfying sleep or experiencing sleep abnormalities were utilizing consumer sleep-tracking devices before seeking professional help from sleep therapists. In the study, two patients were not satisfied with the findings or recommendations of the sleep therapists and did not return for a follow-up visit, citing that the recommendations of the sleep therapists did not correspond with the findings of the consumer sleep-tracking device. The one patient who did return for follow-up visits and followed the recommendations of the sleep therapists, which included switching his device from sensitive mode to normal mode and decreasing his hypnotic medication, was reportedly pleased with the progress made in his sleep quality.

2.4 DIY BIOLOGY AND GENETIC BIOHACKING

2.4.1 CRISPR Kits and At-Home Gene Editing

The discovery of CRISPR-Cas9 technology has revolutionized the field of genetics, opening new possibilities for editing DNA with unprecedented precision and accessibility. What was once the domain of highly advanced laboratories has gradually moved toward more public spaces, including classrooms, citizen science initiatives, and even hobbyist experimentation at home.

This shift has been facilitated by the availability of CRISPR kits, which package the essential tools for gene editing into accessible educational resources. While these kits are not designed to enable advanced genetic engineering of humans or complex organisms, they represent a significant milestone in making biotechnology more approachable for the public. The rise of at-home gene editing is both exciting and controversial, raising questions about education, ethics, safety, and the future of biotechnology.

Basics of CRISPR-Cas9 Technology

CRISPR, short for Clustered Regularly Interspaced Short Palindromic Repeats, is a natural defense mechanism used by bacteria to protect themselves from viral infections. Scientists discovered that this system could be harnessed to edit DNA by using a guiding RNA to target specific genetic sequences, where the Cas9 enzyme acts like molecular scissors to cut and modify the DNA strand. The precision, efficiency, and relative simplicity of this method set it apart from older genetic engineering tools. With CRISPR, researchers can disable genes, insert new sequences, or correct mutations with remarkable ease. This capability has transformed genetic research, agriculture, medicine, and now, through CRISPR kits, education and citizen science.

What Are CRISPR Kits?

CRISPR kits are commercial products that provide the basic materials and instructions needed to perform simple gene-editing experiments. Typically marketed for educational use, these kits allow users to edit the DNA of bacteria or yeast, organisms that are safe to handle and widely used in classroom labs. A common experiment involves editing a gene in bacteria that changes the color of a colony, providing a visible demonstration of CRISPR's power. Some kits include pre-prepared cells, CRISPR components such as guide RNA and Cas9 enzyme, and the necessary nutrients and tools for growing organisms. They are designed to be safe, low-risk, and accessible, often accompanied by detailed manuals or online tutorials. By simplifying the process, CRISPR kits introduce students and hobbyists to biotechnology in a hands-on and engaging way.

Educational Value of CRISPR Kits

One of the main purposes of CRISPR kits is to make genetic science more understandable and interactive. In schools and universities, these kits provide students with the opportunity to see abstract genetic concepts in action. Rather than reading about DNA editing, students can perform simple modifications themselves, bridging the gap between theory and practice. This type of experiential learning inspires curiosity, builds scientific literacy, and can motivate future careers in biotechnology. Beyond formal education, citizen science groups and hobbyist communities use these kits to explore genetics independently. By democratizing access to CRISPR, these kits expand the conversation about biotechnology beyond experts, fostering a more scientifically informed public.

Citizen Science and DIY Biohacking

The availability of CRISPR kits has also intersected with the growing DIY biology and biohacking movements. Community labs around the world offer spaces where non-professionals can experiment with genetic technology in safe, regulated environments. At-home use of CRISPR kits is an extension of this trend, allowing individuals to explore science in their own spaces. Some hobbyists view this as a way to challenge traditional barriers to scientific research, emphasizing open access and collaborative innovation. While the kits are limited in scope and cannot produce advanced genetic modifications, they symbolize a broader shift in how biotechnology is perceived and practiced. The idea that anyone with curiosity and a kit can experiment with DNA reflects both the promise and the unease of democratized science.

Safety Considerations in At-Home Gene Editing

While CRISPR kits are generally designed with safety in mind, questions about the risks of at-home gene editing remain central to the debate. Most commercially available kits use non-pathogenic organisms that do not pose risks to human health or the environment. The experiments are typically simple and educational, with no potential for creating harmful biological products. However, concerns persist that greater accessibility could eventually lead to misuse, either unintentionally through ignorance or deliberately by those seeking to exploit biotechnology. Regulatory frameworks are still evolving to balance the encouragement of citizen science with the need to prevent biohazards. Safety training, guidelines, and community oversight are therefore critical to ensuring that at-home CRISPR experimentation remains beneficial and responsible.

Ethical Dimensions of At-Home Gene Editing

The spread of CRISPR kits into homes and community spaces has also raised important ethical questions. On one hand, democratizing science is seen as a step toward inclusivity and empowerment, giving more people the tools to understand and shape the biological world. On the other hand, the prospect of widespread access to genetic editing raises concerns about bioethics, unintended consequences, and the potential misuse of technology. Some critics worry that making gene editing seem simple and accessible could trivialize its power and complexity, potentially leading to reckless experimentation. Others argue that openness and transparency are essential for public trust in biotechnology. Ethical debates about at-home gene editing therefore reflect broader concerns about who should control access to powerful scientific tools and how responsibility should be shared.

Legal and Regulatory Landscape

Legal approaches to at-home gene editing vary widely across countries. In some places, DIY biology is encouraged as long as safety standards are observed, while in others, strict restrictions are placed on genetic experimentation outside formal institutions. CRISPR kits generally fall into a gray area, as they are marketed for educational purposes and designed to be safe. Nevertheless, regulators keep a close watch on the field, mindful of potential risks and the need for oversight. The challenge lies in creating policies that support innovation and education without opening the door to biosecurity threats. The future of CRISPR kits and at-home gene editing will depend in part on how effectively governments, scientists, and communities navigate this balance.

2.4.2 Synthetic Biology Projects

Synthetic biology represents a fundamental shift in humanity's relationship with the natural world, moving from observation and manipulation to outright design and construction. It is an interdisciplinary field that applies engineering principles to biology, aiming to design and build novel biological entities—such as enzymes, genetic circuits, and cells—or to redesign existing biological systems for useful purposes. Unlike traditional genetic engineering, which often involves transferring a single gene between organisms, synthetic biology envisions assembling comprehensive, standardized genetic modules like electronic components on a circuit board to program organisms with new, predictable functions. This approach transforms biology into a true engineering discipline, with the goal of creating solutions to some of the world's most pressing challenges in medicine, manufacturing, agriculture, and environmental sustainability. The projects emerging from this field are not merely scientific experiments; they are bold demonstrations of biological possibility, blurring the line between the evolved and the invented.

Foundational Tools and Principles: The Basis of Construction

The ambitious goals of synthetic biology are made possible by a suite of foundational technologies that have advanced exponentially over the past two decades. At the core is DNA synthesis, the ability to write genetic code from scratch on a computer and then have it chemically printed, nucleotide by nucleotide. This allows researchers to bypass the constraints of existing natural sequences and create entirely new genetic instructions. Coupled with this is CRISPR-Cas9 gene-editing technology, which acts as a precise pair of molecular scissors, allowing for targeted cuts and edits within a genome with unprecedented accuracy and ease.

These tools are applied within a framework of standardization and modularity known as the BioBrick™ standard. This concept, pioneered at MIT, treats biological parts like resistors and capacitors in an electrical circuit. Standardized DNA sequences, each encoding a specific function (e.g., a promoter to start transcription, a gene for a fluorescent protein, a terminator to stop), can be physically assembled in a modular fashion to create complex genetic circuits. This engineering mindset, combined with rapid drops in the cost of DNA reading and writing, has democratized the field, allowing university labs, startups, and even community biohackers to embark on sophisticated synthetic biology projects.

Medical Marvels: Engineering Biology for Health

One of the most profound applications of synthetic biology is in the realm of medicine, where projects are focused on reprogramming life to diagnose, treat, and prevent disease. A landmark achievement in this area is the microbial production of artemisinin, a potent anti-malarial compound. Traditionally extracted from the sweet wormwood plant in a slow and costly process, synthetic biologists successfully engineered the artemisinin pathway into the yeast Saccharomyces cerevisiae.

By inserting genes from the plant, along with regulatory elements to control their expression, they created a microbial factory that can efficiently produce the drug precursor in large fermenters, ensuring a stable and affordable supply for millions. Beyond pharmaceuticals, synthetic biology is paving the way for advanced cell-based therapies. Researchers are designing immune cells, such as T-cells, with synthetic receptors—CARs (Chimeric Antigen Receptors)—that can recognize and destroy cancer cells with high specificity. These living drugs represent a new frontier in oncology. Furthermore, synthetic biologists are developing sophisticated biosensors: engineered bacteria or mammalian cells that can detect disease biomarkers in a patient sample and produce a visible signal, like fluorescence, enabling rapid and low-cost diagnostics for conditions ranging from infections to cancer.

Sustainable Solutions: Addressing Environmental and Industrial Challenges

The potential of synthetic biology extends far beyond the clinic, offering innovative and sustainable alternatives to traditional industrial processes that are often reliant on petrochemicals and environmentally damaging practices. A major focus is on the creation of biofuels. Numerous projects have engineered microorganisms, particularly bacteria and algae, to efficiently convert sunlight and carbon dioxide, or agricultural waste, into next-generation biofuels like biodiesel, isobutanol, and hydrogen. These renewable energy sources aim to reduce dependence on fossil fuels and lower greenhouse gas emissions. Similarly, the field is revolutionizing materials science. Companies are using engineered microbes to brew sustainable alternatives to leather, silk, and plastics. For instance, by inserting spider silk genes into bacteria or yeast, scientists can produce these incredibly strong and flexible protein fibers in fermenters, avoiding the need for farming spiders. Other projects target environmental remediation, designing organisms capable of detecting and breaking down pollutants like heavy metals, pesticides, and plastic waste in soil and water, effectively using biology to clean up the mess created by human industry.

DIY Movement and Artistic Explorations

Perhaps one of the most fascinating developments in synthetic biology is its democratization through the Do-It-Yourself Biology (DIYbio) movement. Community laboratories, such as Genspace in New York City, provide citizen scientists with access to basic lab equipment and training, enabling them to undertake their own synthetic biology projects. These often focus on open-source and artistic applications, such as engineering bacteria to produce vibrant biological pigments for eco-friendly dyes or creating living artworks where fluorescent bacteria glow in patterns under different environmental conditions. The renowned "Glowing Plant" project, initially a Kickstarter campaign, aimed to use synthetic biology to create plants that emit light, potentially offering a natural alternative to electric street lighting. While these projects may lack the immediate commercial impact of medical or industrial applications, they are critically important. They engage the public in a dialogue about the promises and perils of biological engineering, foster creativity and education, and challenge the notion that this powerful technology is the sole domain of large corporations and elite institutions.

2.4.3 Open-source Biolabs

The landscape of biological research and innovation is undergoing a profound transformation with the emergence of open-source biolabs. These community-driven spaces represent a radical departure from traditional institutional science, offering accessible platforms where individuals—regardless of academic or professional credentials—can engage in biological experimentation, genetic engineering, and biotechnology development.

Unlike conventional laboratories confined to universities or corporate research facilities, open-source biolabs operate on principles of transparency, collaboration, and decentralized innovation. This movement is fueled by the decreasing cost of biotechnology tools, the proliferation of open-access scientific literature, and a growing cultural shift toward citizen science. By providing shared resources, mentorship, and collaborative environments, these labs empower hobbyists, entrepreneurs, and independent researchers to explore biological questions, develop novel solutions, and contribute to scientific progress outside traditional academic or industrial frameworks.

Philosophy and Mission of Open-Source Biolabs

At the heart of open-source biolabs lies a philosophy that challenges the exclusivity of scientific research. These spaces are built on the belief that biological innovation should not be restricted to those with institutional affiliations or substantial funding. Instead, they advocate for a more inclusive model where curiosity, creativity, and community-driven problem-solving can flourish. Many open-source biolabs are guided by principles similar to those of the open-source software movement, emphasizing shared knowledge, open-access tools, and collaborative development.

This ethos fosters an environment where experimentation is encouraged, failures are treated as learning opportunities, and discoveries are openly shared rather than patented or commercialized. The mission extends beyond mere scientific exploration; it encompasses education, public engagement with biotechnology, and the democratization of tools that were once accessible only to elite researchers. By lowering barriers to entry, these labs aim to cultivate a new generation of bio-innovators who can address local and global challenges through grassroots scientific inquiry.

Key Technologies and Tools in Open-Source Biolabs

Open-source biolabs are equipped with an array of tools that enable a wide range of biological experiments, from basic microbiology to advanced genetic engineering. Central to these spaces are polymerase chain reaction (PCR) machines, which allow for DNA amplification and analysis, and electrophoresis equipment for visualizing genetic material. Many labs also feature microscopes, centrifuges, and incubators, enabling experiments in cell culture, microbial growth, and protein analysis. The advent of affordable CRISPR gene-editing kits has further expanded the possibilities within these labs, allowing members to explore genetic modifications in organisms like bacteria, yeast, and even plants. Additionally, open-source software and hardware platforms, such as Arduino-based lab equipment and open-access bioinformatics tools, provide cost-effective alternatives to proprietary systems. These technologies, combined with shared protocols and community expertise, create an environment where complex biological experiments can be conducted with relatively modest resources. The accessibility of these tools has been a game-changer, enabling experiments that were once confined to well-funded institutions to be replicated and innovated upon in community settings.

Community and Collaboration in Open-Source Biolabs

One of the most transformative aspects of open-source biolabs is their emphasis on community and collaboration. Unlike traditional research environments, where competition and intellectual property concerns can limit information sharing, these labs thrive on collective knowledge and peer-to-peer learning. Members often include a diverse mix of scientists, artists, engineers, and entrepreneurs, each bringing unique perspectives to biological challenges. Workshops, hackathons, and collaborative projects are common, fostering an environment where ideas are freely exchanged and refined through group input. Many labs also engage in public outreach, offering educational programs for students, hosting community events, and collaborating with local organizations to address real-world problems. This collaborative spirit extends globally, with open-source biolabs often participating in international networks, sharing protocols, and contributing to open-access databases. The result is a dynamic ecosystem where innovation is accelerated through shared effort rather than siloed research.

Challenges and Ethical Considerations

Despite their innovative potential, open-source biolabs face significant challenges and ethical considerations. One of the primary concerns is biosafety, as the decentralized nature of these labs can lead to inconsistent adherence to safety protocols. While many labs implement rigorous safety measures, the lack of standardized oversight raises questions about the potential risks of accidental contamination or the unintended release of modified organisms. Ethical dilemmas also arise in the context of genetic experimentation, particularly with technologies like CRISPR, where the implications of gene editing extend beyond the lab. The open-source model, while fostering innovation, can sometimes blur the lines between responsible research and reckless experimentation, especially when projects are undertaken without sufficient expertise or oversight. Additionally, the commercialization of discoveries made in these labs can create tensions between the open-source ethos and the realities of intellectual property in a market-driven world. Navigating these challenges requires a delicate balance between fostering innovation and ensuring that experimentation remains safe, ethical, and beneficial to society at large.

Notable Open-Source Biolabs and their Impact

Several open-source biolabs have gained recognition for their contributions to citizen science and biotechnology innovation. Genspace in New York, one of the first community biolabs, has been instrumental in pioneering the open-source biology movement, offering workshops, research opportunities, and public engagement initiatives. Similarly, BioCurious in California has become a hub for bio-innovation, providing resources for entrepreneurs and researchers to develop projects ranging from bioart to synthetic biology.

In Europe, labs like La Paillasse in Paris and Open BioLabs in Belgium have fostered collaborative research environments, often focusing on sustainable and socially impactful projects. These labs have not only advanced scientific knowledge but have also inspired the creation of similar spaces worldwide, demonstrating the global appeal and potential of open-source biology. Their work has contributed to breakthroughs in areas such as environmental monitoring, personalized medicine, and even space biology, showcasing the diverse applications of community-driven science.

2.5 IMPLANTABLE AND ELECTIVE TECHNOLOGY BIOHACKS

2.5.1 RFID/NFC Implants

The merging of biology and technology has given rise to numerous innovations in the field of biohacking, one of the most prominent being RFID and NFC implants. These small electronic devices, inserted under the skin, use radio-frequency identification (RFID) or near-field communication (NFC) technology to interact with external systems. What was once confined to access cards, payment systems, and transportation tickets is now being integrated into the human body as a form of convenience, identity verification, and self-expression. RFID and NFC implants blur the boundary between human and machine, raising questions about utility, safety, ethics, and the future of human augmentation.

RFID and NFC Technology

RFID and NFC are both technologies that use electromagnetic fields to enable communication between a tag and a reader. RFID, which has existed for decades, is widely used in supply chain management, inventory tracking, and security systems. It can transmit data over short to medium ranges depending on frequency. NFC, on the other hand, is a more recent development derived from RFID that allows for close-range communication, typically within a few centimeters. NFC is used in contactless payment systems, digital tickets, and smartphones. Both technologies rely on small chips and antennas that store data and transmit it when activated by a compatible reader. When these technologies are embedded in the human body, they essentially turn people into walking access keys or digital storage devices.

Rise of Biohacking and Implants

RFID and NFC implants emerged from the broader biohacking movement, where individuals experiment with their biology to enhance capabilities. Early adopters, often called grinders, began implanting RFID chips in their hands or arms to replace keys and identification cards. These experiments were initially niche and underground, carried out by enthusiasts willing to test the boundaries of body modification. Over time, however, the concept gained traction, and companies began to commercialize implants for wider audiences. Today, RFID and NFC implants are available for purchase online, marketed as convenient tools for digital interaction. The growing popularity of these implants reflects the increasing cultural acceptance of body-integrated technology.

Practical Uses of RFID and NFC Implants

The primary appeal of RFID and NFC implants lies in their convenience and versatility. One of the most common uses is access control, where an implant replaces traditional keys, fobs, or ID cards for unlocking doors and gates. Similarly, implants can be programmed to store personal information, such as business card details, which can be shared simply by scanning the hand with a smartphone. In some cases, NFC implants have been linked to payment systems, allowing individuals to make purchases with a wave of their hand. Transportation systems, such as subways and buses, have also been accessed through implanted chips in experimental programs. These implants symbolize a seamless integration of technology into daily life, removing the need for external devices such as wallets, keys, or cards.

Cultural and Symbolic Significance

Beyond practical utility, RFID and NFC implants carry symbolic meaning. For some, they represent a form of digital identity, signaling participation in a future-oriented, technologically integrated lifestyle. Biohackers often see implants as a step toward transhumanism, the philosophy that humans can transcend biological limits through technology. The act of implanting a chip is also a statement of self-ownership and autonomy over one's body, challenging traditional norms about human identity. In certain subcultures, implants are viewed as a form of art or expression, no different from tattoos or piercings, but with functional value. This cultural layer makes RFID and NFC implants not only practical tools but also markers of belonging to a growing community of body-technology enthusiasts.

Safety and Medical Considerations

Despite their appeal, RFID and NFC implants raise important safety concerns. The procedure itself is relatively simple, often performed with a syringe in a manner similar to microchipping pets, but it must be done under sterile conditions to prevent infection. Long-term risks include potential migration of the chip within the body or irritation of surrounding tissue. While the chips themselves are passive and do not require power sources, concerns remain about their durability and potential for malfunction. Removal of implants, though possible, can be more complex than insertion. From a medical perspective, these risks are generally considered low, but critics argue that normalization of body implants should not downplay potential complications. Responsible implantation, preferably under professional supervision, is essential for minimizing risks.

Privacy and Security Concerns

Perhaps the most significant issues associated with RFID and NFC implants are privacy and security. Because these devices transmit data wirelessly, they can potentially be read by unauthorized scanners. This raises concerns about hacking, identity theft, and surveillance. Although most consumer-grade implants have limited data capacity and short-range communication, the possibility of malicious use cannot be ignored. As implants begin to interact with financial and identification systems, securing them against unauthorized access becomes increasingly critical. Furthermore, questions of data ownership and consent arise: who controls the information stored in an implant, and how is it protected? These concerns highlight the need for robust encryption and ethical frameworks around body-integrated technology.

Ethical and Societal Debates

The implantation of RFID and NFC devices also fuels ethical debates about autonomy, freedom, and the future of humanity. Supporters argue that implants are voluntary choices that empower individuals with new capabilities and convenience. Critics, however, worry about potential coercion in workplaces, governments, or institutions where implants might be encouraged or even required for access or identification. This raises fears of a surveillance society where bodily integration with technology becomes mandatory. Ethical questions also extend to inequality, as access to body implants could deepen divisions between those who embrace biohacking and those who reject it. The societal conversation about implants is therefore not just about technology but also about values, rights, and human identity.

Role of Companies and Institutions

Commercial interest in RFID and NFC implants has grown steadily, with companies experimenting with their use in offices and workplaces. In some cases, employees have been offered the option to receive implants to access buildings, log into computers, or use vending machines. While framed as a convenience, these programs raise concerns about voluntariness and workplace pressure. At the same time, entrepreneurial companies have emerged to design, sell, and support implants for consumers, often positioning themselves within the biohacking community. The involvement of institutions and businesses suggests that implants may move from niche experimentation to mainstream adoption, depending on social acceptance and regulatory frameworks.

2.5.2 Magnet Implants and Sensory Enhancements

The frontier of biohacking is not limited to compounds and code; it extends into the physical integration of technology with the human body, creating a new class of cyborgs who seek to transcend innate biological limitations. Among the most provocative and tangible practices within this sphere is the implantation of magnets into the body, specifically in the fingertips.

This procedure, once a fringe experiment within the "grinder" subculture, represents a fundamental ambition of the movement: to expand human sensory perception and create a direct, intimate interface with the electromagnetic world that permeates our modern existence. It is a deliberate step towards a future where the boundaries between the organic self and the technological environment become blurred, granting the individual a new sense that is entirely unique to our species.

Procedure and the Principle: Feeling the Invisible

The typical magnet implant involves a small, disc or rod-shaped neodymium magnet that is encased in a biologically inert material such as parylene or silicon nitride to prevent corrosion and rejection by the body's immune system. The procedure is performed under local anesthetic, often by a body modification artist with experience in advanced techniques, as the medical establishment largely does not endorse it. The magnet is carefully inserted into the fingertip, usually of the non-dominant hand, within the dermal layer or just above the connective tissue surrounding the bone.

Over a healing period of several weeks, the body forms a fibrous capsule around the implant, securing it in place. The fundamental principle behind this enhancement is the magnet's ability to transduce electromagnetic energy into a tangible physical sensation. The implant vibrates or exerts a subtle pressure on the nerve-rich fingertip in response to fluctuating electromagnetic fields, effectively allowing the wearer to "feel" these invisible forces. This is not a new sense in the strict neurological definition, but rather a novel form of sensory substitution—a repurposing of the brain's sophisticated tactile interpretation capabilities to perceive a new dimension of reality.

Experiential World: A Sixth Sense in Practice

For those with magnet implants, the world takes on an entirely new layer of texture and information. The most commonly reported experiences involve interacting with everyday electronics. A hairdryer, microwave oven, or power transformer generates a strong, humming sensation, revealing the intensity of the field it emits. Walking through a space can reveal the hidden wiring within walls as the magnet responds to the alternating current flowing through it. The most profound interaction is often with devices containing speakers, as the magnet allows the wearer to feel the vibrations of the audio waveform itself, creating a direct tactile connection to music and sound. Beyond mere novelty, this new perception can have practical applications. Electricians and engineers report an ability to diagnose faulty wiring or malfunctioning motors by the aberrant or absent fields they produce. It creates a constant, silent dialogue between the individual and the manufactured world, fostering an intuitive understanding of the energy flows that power modern civilization. The sensation is frequently described as a low-grade vibration, a subtle pull, or a pressure that is felt rather than heard, a private and continuous reminder of the unseen forces that surround us.

Beyond Magnets: The Expanding Horizon of Sensory Augmentation

While magnet implants are the most established form of sensory enhancement, they represent just the beginning of a broader exploration into expanding human perception. The same grinder community that pioneered magnetic implants has experimented with other devices, such as NorthSense, a small implant that vibrates when facing magnetic north, effectively granting a constant, internalized sense of direction. Other projects explore the boundaries of sensory input by attempting to perceive other forms of energy. Experimental implants include RFID (Radio-Frequency Identification) and NFC (Near-Field Communication) chips, which are typically implanted in the hand. While often used for practical tasks like unlocking doors or storing digital business cards, these chips also represent a form of sensory enhancement, allowing the wearer to interact with and sense the proximity of RFID readers in their environment. More speculative research points towards future enhancements that could allow humans to perceive ultraviolet or infrared light, sense ionizing radiation directly, or even develop an interface for perceiving real-time data streams from the internet, translating digital information into somatic sensations.

Body as a Platform: Philosophical and Ethical Dimensions

The decision to undergo a magnet implant is not merely a technical one; it is deeply philosophical. It embodies a core tenet of transhumanism: that the human body is not a sacred, finished product but a malleable platform that can and should be improved upon. It is a declaration of ownership over one's own biology and a rejection of the notion that our senses provide a complete or accurate picture of reality. By adding a new sense, the biohacker actively participates in their own evolution, choosing which aspects of the universe they wish to perceive. However, this practice is fraught with significant ethical and safety considerations.

The procedure carries risks of infection, nerve damage, and long-term rejection. The magnetic field itself could potentially interfere with medical devices like MRIs, presenting a life-threatening situation if not disclosed. Furthermore, the largely unregulated nature of the practice means that individuals rely on community knowledge and the skill of specific practitioners, operating outside the safeguards of formal medical oversight. This raises questions about informed consent, the long-term health effects of chronic exposure to electromagnetic fields on nervous tissue, and the societal implications of a future where sensory augmentation could create a new form of inequality between the enhanced and the non-enhanced.

2.5.3 Brain-Computer Interfaces (BCIs)

A brain–computer interface (BCI), sometimes called a brain–machine interface (BMI), is a direct communication link between the brain's electrical activity and an external device, most commonly a computer or robotic limb. BCIs are often directed at researching, mapping, assisting, augmenting, or repairing human cognitive or sensory-motor functions. They are often conceptualized as a human–machine interface that skips the intermediary of moving body parts (e.g. hands or feet). BCI implementations range from non-invasive (EEG, MEG, MRI) and partially invasive (ECoG and endovascular) to invasive (microelectrode array), based on how physically close electrodes are to brain tissue.

Research on BCIs began in the 1970s by Jacques Vidal at the University of California, Los Angeles (UCLA) under a grant from the National Science Foundation, followed by a contract from the Defense Advanced Research Projects Agency (DARPA). Vidal's 1973 paper introduced the expression brain–computer interface into scientific literature. Due to the cortical plasticity of the brain, signals from implanted prostheses can, after adaptation, be handled by the brain like natural sensor or effector channels. Following years of animal experimentation, the first neuroprosthetic devices were implanted in humans in the mid-1990s.

History

The history of brain-computer interfaces (BCIs) starts with Hans Berger's discovery of the brain's electrical activity and the development of electroencephalography (EEG). In 1924 Berger was the first to record human brain activity utilizing EEG.

Berger was able to identify oscillatory activity, such as the alpha wave (8–13 Hz), by analyzing EEG traces. Berger's first recording device was rudimentary. He inserted silver wires under the scalps of his patients. These were later replaced by silver foils attached to the patient's head by rubber bandages. Berger connected these sensors to a Lippmann capillary electrometer, with disappointing results. However, more sophisticated measuring devices, such as the Siemens double-coil recording galvanometer, which displayed voltages as small as 10^{-4} volt, led to success. Berger analyzed the interrelation of alternations in his EEG wave diagrams with brain diseases. EEGs permitted completely new possibilities for brain research.

Although the term had not yet been coined, one of the earliest examples of a working brain-machine interface was the piece Music for Solo Performer (1965) by American composer Alvin Lucier. The piece makes use of EEG and analog signal processing hardware (filters, amplifiers, and a mixing board) to stimulate acoustic percussion instruments. Performing the piece requires producing alpha waves and thereby "playing" the various instruments via loudspeakers that are placed near or directly on the instruments.

Jacques Vidal coined the term "BCI" and produced the first peer-reviewed publications on this topic. He is widely recognized as the inventor of BCIs. A review pointed out that Vidal's 1973 paper stated the "BCI challenge" of controlling external objects using EEG signals, and especially use of Contingent Negative Variation (CNV) potential as a challenge for BCI control. Vidal's 1977 experiment was the first application of BCI after his 1973 BCI challenge. It was a noninvasive EEG (actually Visual Evoked Potentials (VEP)) control of a cursor-like graphical object on a computer screen. The demonstration was movement in a maze.

1988 was the first demonstration of noninvasive EEG control of a physical object, a robot. The experiment demonstrated EEG control of multiple start-stop-restart cycles of movement, along an arbitrary trajectory defined by a line drawn on a floor. The line-following behavior was the default robot behavior, utilizing autonomous intelligence and an autonomous energy source.

In 1990, a report was given on a closed loop, bidirectional, adaptive BCI controlling a computer buzzer by an anticipatory brain potential, the Contingent Negative Variation (CNV) potential. The experiment described how an expectation state of the brain, manifested by CNV, used a feedback loop to control the S2 buzzer in the S1-S2-CNV paradigm. The resulting cognitive wave representing the expectation learning in the brain was termed Electroexpectogram (EXG). The CNV brain potential was part of Vidal's 1973 challenge. Studies in the 2010s suggested neural stimulation's potential to restore functional connectivity and associated behaviors through modulation of molecular mechanisms. This opened the door for the concept that BCI technologies may be able to restore function. Beginning in 2013,

DARPA funded BCI technology through the BRAIN initiative, which supported work out of teams including University of Pittsburgh Medical Center, Paradromics, Brown, and Synchron.

Neuroprosthetics

Neuroprosthetics is an area of neuroscience concerned with neural prostheses, that is, using artificial devices to replace the function of impaired nervous systems and brain-related problems, or of sensory or other organs (bladder, diaphragm, etc.). As of December 2010, cochlear implants had been implanted as neuroprosthetic devices in some 736,900 people worldwide. Other neuroprosthetic devices aim to restore vision, including retinal implants. The first neuroprosthetic device, however, was the pacemaker. The terms are sometimes used interchangeably. Neuroprosthetics and BCIs seek to achieve the same aims, such as restoring sight, hearing, movement, ability to communicate, and even cognitive function. Both use similar experimental methods and surgical techniques.

Animal Research

Several laboratories have managed to read signals from monkey and rat cerebral cortices to operate BCIs to produce movement. Monkeys have moved computer cursors and commanded robotic arms to perform simple tasks simply by thinking about the task and seeing the results, without motor output. In May 2008 photographs that showed a monkey at the University of Pittsburgh Medical Center operating a robotic arm by thinking were published in multiple studies. Sheep have also been used to evaluate BCI technology including Synchron's Stentrode. In 2020, Elon Musk's Neuralink was successfully implanted in a pig. In 2021, Musk announced that the company had successfully enabled a monkey to play video games using Neuralink's device.

Early Work

In 1969 operant conditioning studies by Fetz et al. at the Regional Primate Research Center and Department of Physiology and Biophysics, University of Washington School of Medicine showed that monkeys could learn to control the deflection of a biofeedback arm with neural activity. Similar work in the 1970s established that monkeys could learn to control the firing rates of individual and multiple neurons in the primary motor cortex if they were rewarded accordingly.

Algorithms to reconstruct movements from motor cortex neurons, which control movement, date back to the 1970s. In the 1980s, Georgopoulos at Johns Hopkins University found a mathematical relationship between the electrical responses of single motor cortex neurons in rhesus macaque monkeys and the direction in which they moved their arms. He also found that dispersed groups of neurons, in different areas of the monkey's brains, collectively controlled motor commands.

He was able to record the firings of neurons in only one area at a time, due to equipment limitations. Several groups have been able to capture complex brain motor cortex signals by recording from neural ensembles (groups of neurons) and using these to control external devices.

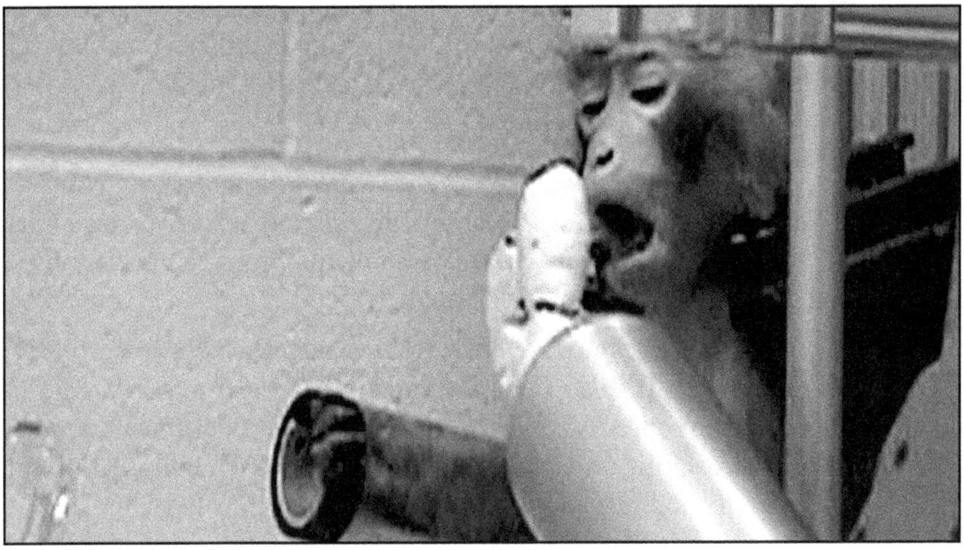

Monkey Operating a Robotic Arm with Brain—Computer Interfacing

Research

Kennedy and Yang Dan:

- Phillip Kennedy (Neural Signals founder (1987) and colleagues built the first intracortical brain-computer interface by implanting neurotrophic-cone electrodes into monkeys.

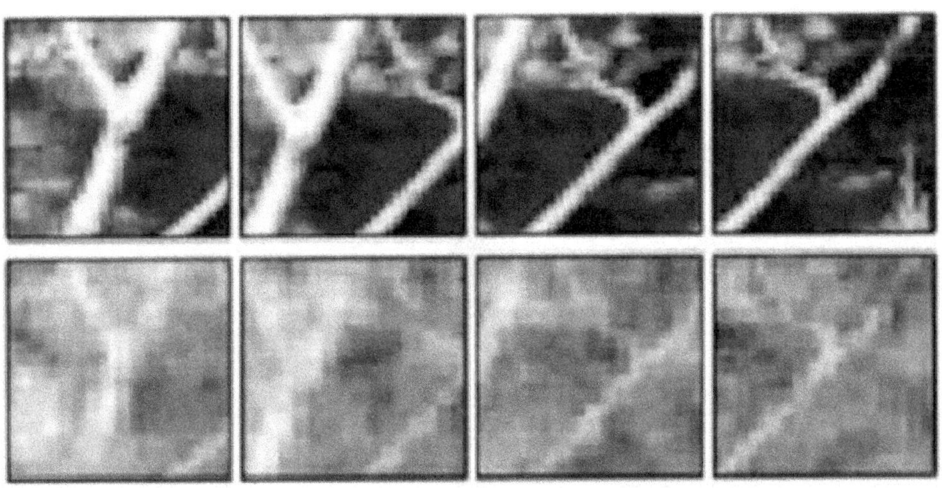

Yang Dan and Colleagues' Recordings of Cat Vision Using a BCI Implanted in the Lateral Geniculate Nucleus (Top Row—Original Image; Bottom Row—Recording)

- In 1999, Yang Dan et al. at University of California, Berkeley decoded neuronal firings to reproduce images from cats. The team used an array of electrodes embedded in the thalamus (which integrates the brain's sensory input). Researchers targeted 177 brain cells in the thalamus lateral geniculate nucleus area, which decodes signals from the retina. Neuron firings were recorded from watching eight short movies. Using mathematical filters, the researchers decoded the signals to reconstruct recognizable scenes and moving objects.

Nicolelis:

- Duke University professor Miguel Nicolelis advocates using multiple electrodes spread over a greater area of the brain to obtain neuronal signals.
- After initial studies in rats during the 1990s, Nicolelis and colleagues developed BCIs that decoded brain activity in owl monkeys and used the devices to reproduce monkey movements in robotic arms. Monkeys' advanced reaching and grasping abilities and hand manipulation skills, made them good test subjects.

By 2000, the group succeeded in building a BCI that reproduced owl monkey movements while the monkey operated a joystick or reached for food. The BCI operated in real time and could remotely control a separate robot. But the monkeys received no feedback (open-loop BCI).

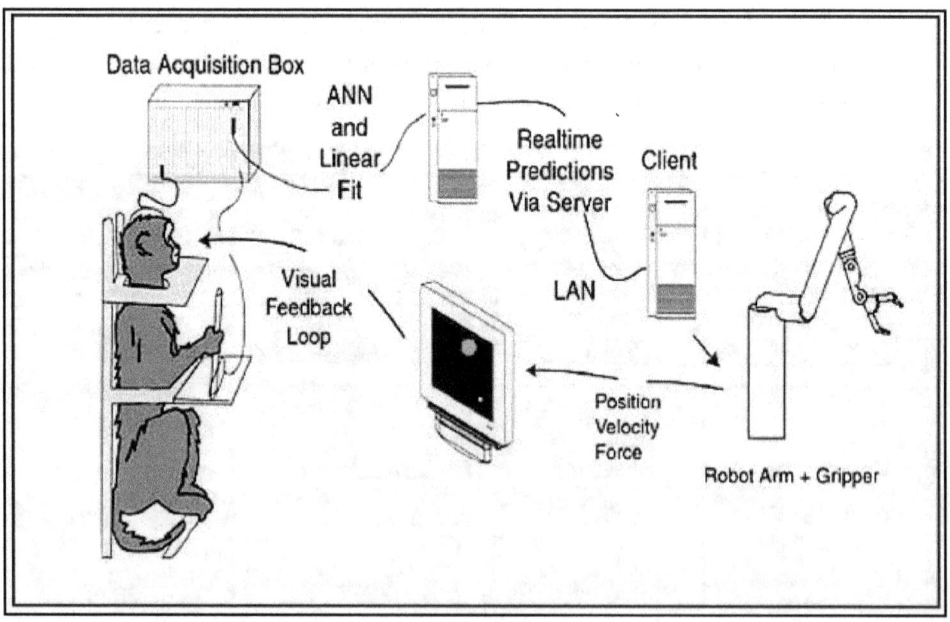

Diagram of the BCI Developed by Miguel Nicolelis and Colleagues for Use on Rhesus Monkeys

- Later experiments on rhesus monkeys included feedback and reproduced monkey reaching and grasping movements in a robot arm. Their deeply cleft and furrowed brains made them better models for human neurophysiology than owl monkeys. The monkeys were trained to reach and grasp objects on a computer screen by manipulating a joystick while corresponding movements by a robot arm were hidden. The monkeys were later shown the robot and learned to control it by viewing its movements. The BCI used velocity predictions to control reaching movements and simultaneously predicted gripping force.
- In 2011 O'Doherty and colleagues showed a BCI with sensory feedback with rhesus monkeys. The monkey controlled the position of an avatar arm while receiving sensory feedback through direct intracortical stimulation (ICMS) in the arm representation area of the sensory cortex.

Donoghue, Schwartz, and Andersen:

- Other laboratories that have developed BCIs and algorithms that decode neuron signals include John Donoghue at the Carney Institute for Brain Science at Brown University, Andrew Schwartz at the University of Pittsburgh, and Richard Andersen at Caltech. These researchers produced working BCIs using recorded signals from far fewer neurons than Nicolelis (15–30 neurons versus 50–200 neurons).
- The Carney Institute reported training rhesus monkeys to use a BCI to track visual targets on a computer screen (closed-loop BCI) with or without a joystick. The group created a BCI for three-dimensional tracking in virtual reality and reproduced BCI control in a robotic arm. The same group demonstrated that a monkey could feed itself pieces of fruit and marshmallows using a robotic arm controlled by the animal's brain signals.
- Andersen's group used recordings of premovement activity from the posterior parietal cortex, including signals created when experimental animals anticipated receiving a reward.

Other Research:

- In addition to predicting kinematic and kinetic parameters of limb movements, BCIs that predict electromyographic or electrical activity of the muscles of primates are in process. Such BCIs could restore mobility in paralyzed limbs by electrically stimulating muscles.
- Nicolelis and colleagues demonstrated that large neural ensembles can predict arm position. This work allowed BCIs to read arm movement intentions and translate them into actuator movements. Carmena and

colleagues programmed a BCI that allowed a monkey to control reaching and grasping movements by a robotic arm. Lebedev and colleagues argued that brain networks reorganize to create a new representation of the robotic appendage in addition to the representation of the animal's own limbs.

- In 2019, researchers from the University of San Francisco, California, initiated a brain-computer interface (BCI) study that had the potential to aid patients with speech impairment resulting from neurological disorders. Their BCI utilized high-density electrocorticography to capture neural activity from a patient's brain and employed deep learning to synthesize speech. In 2021, those researchers reported the potential of a BCI to decode words and sentences in an anarthric patient who had been unable to speak for over 15 years.
- The biggest impediment to BCI technology is the lack of a sensor modality that provides safe, accurate and robust access to brain signals. The use of a better sensor expands the range of communication functions that can be provided using a BCI.
- Development and implementation of a BCI system is complex and time-consuming. In response to this problem, Gerwin Schalk has been developing BCI2000, a general-purpose system for BCI research, since 2000.
- A new 'wireless' approach uses light-gated ion channels such as channelrhodopsin to control the activity of genetically defined subsets of neurons in vivo. In the context of a simple learning task, illumination of transfected cells in the somatosensory cortex influenced decision-making in mice.
- BCIs led to a deeper understanding of neural networks and the central nervous system. Research has reported that despite neuroscientists' inclination to believe that neurons have the most effect when working together, single neurons can be conditioned through the use of BCIs to fire in a pattern that allows primates to control motor outputs. BCIs led to development of the single neuron insufficiency principle that states that even with a well-tuned firing rate, single neurons can only carry limited information and therefore the highest level of accuracy is achieved by recording ensemble firings. Other principles discovered with BCIs include the neuronal multitasking principle, the neuronal mass principle, the neural degeneracy principle, and the plasticity principle.
- BCIs are proposed to be applied by users without disabilities. Passive BCIs allow for assessing and interpreting changes in the user state

during Human–computer interaction (HCI). In a secondary, implicit control loop, the system adapts to its user, improving its usability.

- BCI systems can potentially be used to encode signals from the periphery. These sensory BCI devices enable real-time, behaviorally-relevant decisions based upon closed-loop neural stimulation.

Human Research

Invasive BCIs

Invasive BCI requires surgery to implant electrodes under the scalp for accessing brain signals. The main advantage is to increase accuracy. Downsides include side effects from the surgery, including scar tissue that can obstruct brain signals, or the body potentially rejecting the implanted electrodes.

Vision:

- Invasive BCI research has targeted repairing damaged sight and providing new functionality for people with paralysis. Invasive BCIs are implanted directly into the grey matter of the brain during neurosurgery. Because they lie in the grey matter, invasive devices produce the highest quality signals of BCI devices but are prone to scar-tissue build-up, causing the signal to weaken, or disappear, as the body reacts to the foreign object.
- In vision science, direct brain implants have been used to treat non-congenital (acquired) blindness. One of the first scientists to produce a working brain interface to restore sight was private researcher William Dobelle. Dobelle's first prototype was implanted into "Jerry", a man blinded in adulthood, in 1978. A single-array BCI containing 68 electrodes was implanted onto Jerry's visual cortex and succeeded in producing phosphenes, the sensation of seeing light. The system included cameras mounted on glasses to send signals to the implant. Initially, the implant allowed Jerry to see shades of grey in a limited field of vision at a low frame-rate. This also required him to be hooked up to a mainframe computer, but shrinking electronics and faster computers made his artificial eye more portable and now enable him to perform simple tasks unassisted.
- In 2002, Jens Naumann, also blinded in adulthood, became the first in a series of 16 paying patients to receive Dobelle's second generation implant, one of the earliest commercial uses of BCIs. The second generation device used a more sophisticated implant enabling better mapping of phosphenes into coherent vision. Phosphenes are spread out across the visual field in what researchers call "the starry-night effect".

Immediately after his implant, Jens was able to use his imperfectly restored vision to drive an automobile slowly around the parking area of the research institute. Dobelle died in 2004 before his processes and developments were documented, leaving no one to continue his work. Subsequently, Naumann and the other patients in the program began having problems with their vision, and eventually lost their "sight" again.

Movement:

- BCIs focusing on motor neuroprosthetics aim to restore movement in individuals with paralysis or provide devices to assist them, such as interfaces with computers or robot arms.
- Kennedy and Bakay were first to install a human brain implant that produced signals of high enough quality to simulate movement. Their patient, Johnny Ray (1944–2002), developed 'locked-in syndrome' after a brain-stem stroke in 1997. Ray's implant was installed in 1998 and he lived long enough to start working with the implant, eventually learning to control a computer cursor; he died in 2002 of a brain aneurysm.
- Tetraplegic Matt Nagle became the first person to control an artificial hand using a BCI in 2005 as part of the first nine-month human trial of Cyberkinetics's BrainGate chip-implant. Implanted in Nagle's right precentral gyrus (area of the motor cortex for arm movement), the 96-electrode implant allowed Nagle to control a robotic arm by thinking about moving his hand as well as a computer cursor, lights and TV. One year later, Jonathan Wolpaw received the Altran Foundation for Innovation prize for developing a Brain Computer Interface with electrodes located on the surface of the skull, instead of directly in the brain.
- Research teams led by the BrainGate group and another at University of Pittsburgh Medical Center, both in collaborations with the United States Department of Veterans Affairs (VA), demonstrated control of prosthetic limbs with many degrees of freedom using direct connections to arrays of neurons in the motor cortex of tetraplegia patients.

Communication:

- In May 2021, a Stanford University team reported a successful proof-of-concept test that enabled a quadraplegic participant to produce English sentences at about 86 characters per minute and 18 words per minute. The participant imagined moving his hand to write letters, and the system performed handwriting recognition on electrical signals detected in the motor cortex, utilizing Hidden Markov models and recurrent neural networks. Since researchers from UCSF initiated a brain-computer interface (BCI) study, numerous reports have been made. In 2021, they reported that a paralyzed and with anarthria man was able to communicate fifteen words per minute using an implanted device that examined nerve cells controlling the muscles of the vocal tract. In addition in 2022 it was announced that their implant could also be used to spell out words and entire sentences without speaking aloud. The first bilingual speech neuroprosthesis was reported to have been developed by the same team at the University of San Francisco, in 2024.

- In a review article, authors wondered whether human information transfer rates can surpass that of language with BCIs. Language research has reported that information transfer rates are relatively constant across many languages. This may reflect the brain's information processing limit. Alternatively, this limit may be intrinsic to language itself, as a modality for information transfer.
- In 2023 two studies used BCIs with recurrent neural network to decode speech at a record rate of 62 words per minute and 78 words per minute.

Technical Challenges:

- There exist a number of technical challenges to recording brain activity with invasive BCIs. Advances in CMOS technology are pushing and enabling integrated, invasive BCI designs with smaller size, lower power requirements, and higher signal acquisition capabilities. Invasive BCIs involve electrodes that penetrate brain tissue in an attempt to record action potential signals (also known as spikes) from individual, or small groups of, neurons near the electrode. The interface between a recording electrode and the electrolytic solution surrounding neurons has been modelled using the Hodgkin-Huxley model.
- Electronic limitations to invasive BCIs have been an active area of research in recent decades. While intracellular recordings of neurons reveal action potential voltages on the scale of hundreds of millivolts, chronic invasive BCIs rely on recording extracellular voltages which typically are three orders of magnitude smaller, existing at hundreds of microvolts. Further adding to the challenge of detecting signals on the scale of microvolts is the fact that the electrode-tissue interface has a high capacitance at small voltages. Due to the nature of these small signals, for BCI systems that incorporate functionality onto an integrated circuit, each electrode requires its own amplifier and ADC, which convert analog extracellular voltages into digital signals.

Because a typical neuron action potential lasts for one millisecond, BCIs measuring spikes must have sampling rates ranging from 300 Hz to 5 kHz. Yet another concern is that invasive BCIs must be low-power, so as to dissipate less heat to surrounding tissue; at the most basic level more power is traditionally needed to optimize signal-to-noise ratio. Optimal battery design is an active area of research in BCIs.

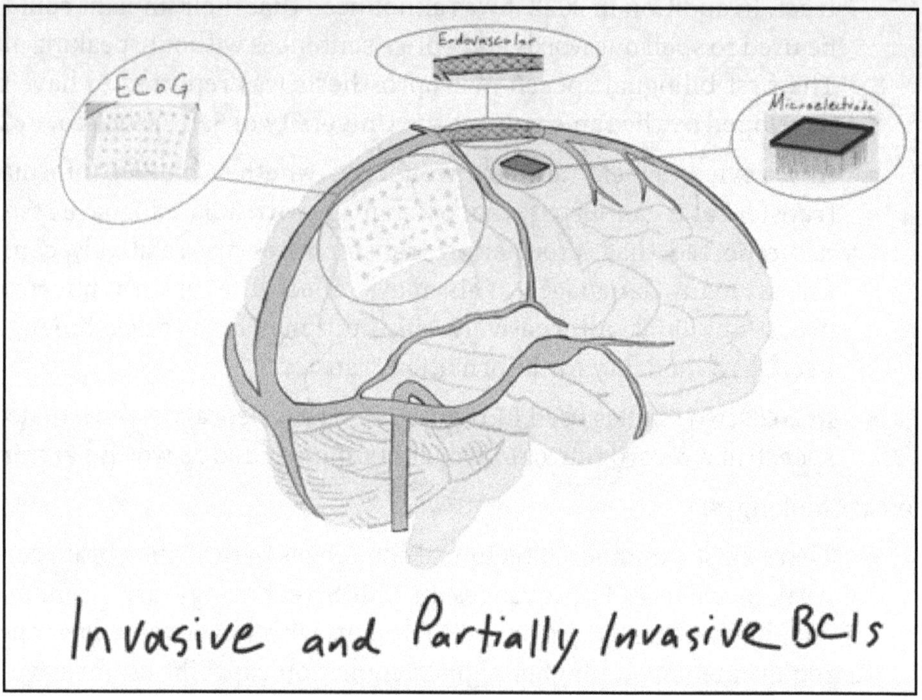

Illustration of Invasive and Partially Invasive BCIs—Electrocorticography (ECoG), Endovascular, and Intracortical Microelectrode

- Challenges existing in the area of material science are central to the design of invasive BCIs. Variations in signal quality over time have been commonly observed with implantable microelectrodes. Optimal material and mechanical characteristics for long term signal stability in invasive BCIs has been an active area of research. It has been proposed that the formation of glial scarring, secondary to damage at the electrode-tissue interface, is likely responsible for electrode failure and reduced recording performance. Research has suggested that blood-brain barrier leakage, either at the time of insertion or over time, may be responsible for the inflammatory and glial reaction to chronic microelectrodes implanted in the brain. As a result, flexible and tissue-like designs have been researched and developed to minimize foreign-body reaction by means of matching the Young's modulus of the electrode closer to that of brain tissue.

Partially Invasive BCIs

Partially invasive BCI devices are implanted inside the skull but rest outside the brain rather than within the grey matter. They produce higher resolution signals than non-invasive BCIs where the bone tissue of the cranium deflects and deforms signals and have a lower risk of forming scar-tissue in the brain than fully invasive BCIs. Preclinical demonstration of intracortical BCIs from the stroke perilesional cortex has been conducted.

Endovascular:

- A systematic review published in 2020 detailed multiple clinical and non-clinical studies investigating the feasibility of endovascular BCIs.
- In 2010, researchers affiliated with University of Melbourne began developing a BCI that could be inserted via the vascular system. Australian neurologist Thomas Oxley conceived the idea for this BCI, called Stentrode, earning funding from DARPA. Preclinical studies evaluated the technology in sheep.
- Stentrode is a monolithic stent electrode array designed to be delivered via an intravenous catheter under image-guidance to the superior sagittal sinus, in the region which lies adjacent to the motor cortex. This proximity enables Stentrode to measure neural activity. The procedure is most similar to how venous sinus stents are placed for the treatment of idiopathic intracranial hypertension. Stentrode communicates neural activity to a battery-less telemetry unit implanted in the chest, which communicates wirelessly with an external telemetry unit capable of power and data transfer. While an endovascular BCI benefits from avoiding a craniotomy for insertion, risks such as clotting and venous thrombosis exist.
- Human trials with Stentrode were underway as of 2021. In November 2020, two participants with amyotrophic lateral sclerosis were able to wirelessly control an operating system to text, email, shop, and bank using direct thought using Stentrode, marking the first time a brain-computer interface was implanted via the patient's blood vessels, eliminating the need for brain surgery. In January 2023, researchers reported no serious adverse events during the first year for all four patients, who could use it to operate computers.

Electrocorticography:

- Electrocorticography (ECoG) measures brain electrical activity from beneath the skull in a way similar to non-invasive electroencephalography, using electrodes embedded in a thin plastic pad placed above the cortex, beneath the dura mater. ECoG technologies were first trialled in humans in 2004 by Eric Leuthardt and Daniel Moran from Washington University in St. Louis. In a later trial, the researchers enabled a teenage boy to play Space Invaders. This research indicates that control is rapid, requires minimal training, balancing signal fidelity and level of invasiveness.

- Signals can be either subdural or epidural, but are not taken from within the brain parenchyma. Patients are required to have invasive monitoring for localization and resection of an epileptogenic focus.
- ECoG offers higher spatial resolution, better signal-to-noise ratio, wider frequency range, and less training requirements than scalp-recorded EEG, and at the same time has lower technical difficulty, lower clinical risk, and may have superior long-term stability than intracortical single-neuron recording. This feature profile and evidence of the high level of control with minimal training requirements shows potential for real world application for people with motor disabilities.
- Edward Chang and Joseph Makin from UCSF reported that ECoG signals could be used to decode speech from epilepsy patients implanted with high-density ECoG arrays over the peri-Sylvian cortices. They reported word error rates of 3% (a marked improvement from prior efforts) utilizing an encoder-decoder neural network, which translated ECoG data into one of fifty sentences composed of 250 unique words.

Functional Near-Infrared Spectroscopy:

- In 2014, a BCI using functional near-infrared spectroscopy for "locked-in" patients with amyotrophic lateral sclerosis (ALS) was able to restore basic ability to communicate.

Electroencephalography (EEG)-Based Brain-Computer Interfaces:

- After Vidal stated the BCI challenge, the initial reports on non-invasive approaches included control of a cursor in 2D using VEP, control of a buzzer using CNV, control of a physical object, a robot, using a brain rhythm (alpha), control of a text written on a screen using P300.
- In the early days of BCI research, another substantial barrier to using EEG was that extensive training was required. For example, in experiments beginning in the mid-1990s, Niels Birbaumer at the University of Tübingen in Germany trained paralysed people to self-regulate the slow cortical potentials in their EEG to such an extent that these signals could be used as a binary signal to control a computer cursor. (Birbaumer had earlier trained epileptics to prevent impending fits by controlling

this low voltage wave.) The experiment trained ten patients to move a computer cursor. The process was slow, requiring more than an hour for patients to write 100 characters with the cursor, while training often took months. The slow cortical potential approach has fallen away in favor of approaches that require little or no training, are faster and more accurate, and work for a greater proportion of users.

- Another research parameter is the type of oscillatory activity that is measured. Gert Pfurtscheller founded the BCI Lab 1991 and conducted the first online BCI based on oscillatory features and classifiers. Together with Birbaumer and Jonathan Wolpaw at New York State University they focused on developing technology that would allow users to choose the brain signals they found easiest to operate a BCI, including mu and beta rhythms.
- A further parameter is the method of feedback used as shown in studies of P300 signals. Patterns of P300 waves are generated involuntarily (stimulus-feedback) when people see something they recognize and may allow BCIs to decode categories of thoughts without training.
- A 2005 study reported EEG emulation of digital control circuits, using a CNV flip-flop. A 2009 study reported noninvasive EEG control of a robotic arm using a CNV flip-flop. A 2011 study reported control of two robotic arms solving Tower of Hanoi task with three disks using a CNV flip-flop. A 2015 study described EEG-emulation of a Schmitt trigger, flip-flop, demultiplexer, and modem.
- Advances by Bin He and his team at University of Minnesota suggest the potential of EEG-based brain-computer interfaces to accomplish tasks close to invasive brain-computer interfaces. Using advanced functional neuroimaging including BOLD functional MRI and EEG source imaging, They identified the co-variation and co-localization of electrophysiological and hemodynamic signals. Refined by a neuroimaging approach and a training protocol, They fashioned a non-invasive EEG based brain-computer interface to control the flight of a virtual helicopter in 3-dimensional space, based upon motor imagination. In June 2013 they announced a technique to guide a remote-control helicopter through an obstacle course. They also solved the EEG inverse problem and then used the resulting virtual EEG for BCI tasks. Well-controlled studies suggested the merits of such a source analysis-based BCI.
- A 2014 study reported that severely motor-impaired patients could communicate faster and more reliably with non-invasive EEG BCI than with muscle-based communication channels.

- A 2019 study reported that the application of evolutionary algorithms could improve EEG mental state classification with a non-invasive Muse device, enabling classification of data acquired by a consumer-grade sensing device.
- In a 2021 systematic review of randomized controlled trials using BCI for post-stroke upper-limb rehabilitation, EEG-based BCI was reported to have efficacy in improving upper-limb motor function compared to control therapies. More specifically, BCI studies that utilized band power features, motor imagery, and functional electrical stimulation were reported to be more effective than alternatives. Another 2021 systematic review focused on post-stroke robot-assisted EEG-based BCI for hand rehabilitation. Improvement in motor assessment scores was observed in three of eleven studies.

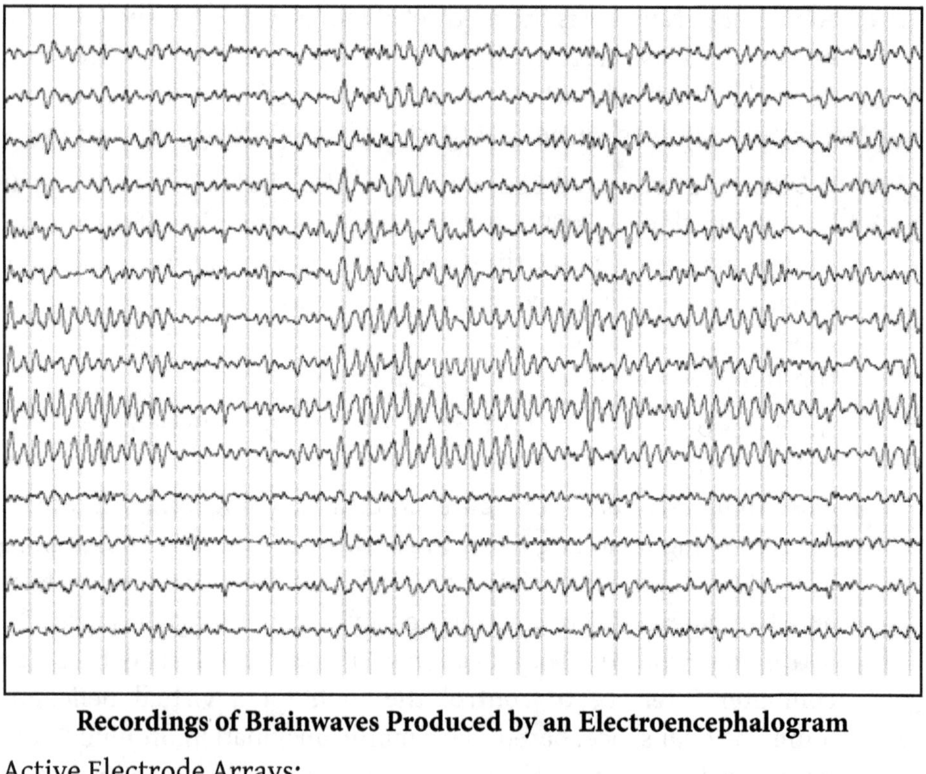

Recordings of Brainwaves Produced by an Electroencephalogram

Dry Active Electrode Arrays:

- In the early 1990s Babak Taheri, at University of California, Davis demonstrated the first single and multichannel dry active electrode arrays. The arrayed electrode was demonstrated to perform well compared to silver/silver chloride electrodes. The device consisted of four sensor sites with integrated electronics to reduce noise by impedance matching. The advantages of such electrodes are:

  - No electrolyte used,
  - No skin preparation,
  - Significantly reduced sensor size,
  - Compatibility with EEG monitoring systems.
- The active electrode array is an integrated system containing an array of capacitive sensors with local integrated circuitry packaged with batteries to power the circuitry. This level of integration was required to achieve the result.
- The electrode was tested on a test bench and on human subjects in four modalities, namely:
  - Spontaneous EEG,
  - Sensory event-related potentials,
  - Brain stem potentials,
  - Cognitive event-related potentials.
- Performance compared favorably with that of standard wet electrodes in terms of skin preparation, no gel requirements (dry), and higher signal-to-noise ratio.
- In 1999 Hunter Peckham and others at Case Western Reserve University used a 64-electrode EEG skullcap to return limited hand movements to a quadriplegic. As he concentrated on simple but opposite concepts like up and down. A basic pattern was identified in his beta-rhythm EEG output and used to control a switch: Above average activity was interpreted as on, below average off. The signals were also used to drive nerve controllers embedded in his hands, restoring some movement.

SSVEP Mobile EEG BCIs:

- In 2009, the NCTU Brain-Computer-Interface-headband was announced. Those researchers also engineered silicon-based microelectro-mechanical system (MEMS) dry electrodes designed for application to non-hairy body sites. These electrodes were secured to the headband's DAQ board with snap-on electrode holders. The signal processing module measured alpha activity and transferred it over Bluetooth to a phone that assessed the patients' alertness and cognitive capacity. When the subject became drowsy, the phone sent arousing feedback to the operator to rouse them.
- In 2011, researchers reported a cellular based BCI that could cause a phone to ring. The wearable system was composed of a four channel bio-signal acquisition/amplification module, a communication module, and a Bluetooth phone. The electrodes were placed to pick up steady

state visual evoked potentials (SSVEPs). SSVEPs are electrical responses to flickering visual stimuli with repetition rates over 6 Hz that are best found in the parietal and occipital scalp regions of the visual cortex. It was reported that all study participants were able to initiate the phone call with minimal practice in natural environments.

- The scientists reported that a single channel fast Fourier transform (FFT) and multiple channel system canonical correlation analysis (CCA) algorithm can support mobile BCIs. The CCA algorithm has been applied in experiments investigating BCIs with claimed high accuracy and speed. Cellular BCI technology can reportedly be translated for other applications, such as picking up sensorimotor mu/beta rhythms to function as a motor-imagery based BCI.
- In 2013, comparative tests performed on Android cell phone, tablet, and computer based BCIs, analyzed the power spectrum density of resultant EEG SSVEPs. The stated goals of this study were to "increase the practicability, portability, and ubiquity of an SSVEP-based BCI, for daily use". It was reported that the stimulation frequency on all mediums was accurate, although the phone's signal was not stable. The amplitudes of the SSVEPs for the laptop and tablet were reported to be larger than those of the cell phone. These two qualitative characterizations were suggested as indicators of the feasibility of using a mobile stimulus BCI.
- One of the difficulties with EEG readings is susceptibility to motion artifacts. In most research projects, the participants were asked to sit still in a laboratory setting, reducing head and eye movements as much as possible. However, since these initiatives were intended to create a mobile device for daily use, the technology had to be tested in motion. In 2013, researchers tested mobile EEG-based BCI technology, measuring SSVEPs from participants as they walked on a treadmill. Reported results were that as speed increased, SSVEP detectability using CCA decreased. Independent component analysis (ICA) had been shown to be efficient in separating EEG signals from noise. The researchers stated that CCA data with and without ICA processing were similar. They concluded that CCA demonstrated robustness to motion artifacts. EEG-based BCI applications offer low spatial resolution. Possible solutions include: EEG source connectivity based on graph theory, EEG pattern recognition based on Topomap and EEG-fMRI fusion.

Prosthesis and Environment Control:

- Non-Invasive BCIs have been applied to prosthetic upper and lower extremity devices in people with paralysis. For example,

Gert Pfurtscheller of Graz University of Technology and colleagues demonstrated a BCI-controlled functional electrical stimulation system to restore upper extremity movements in a person with tetraplegia due to spinal cord injury. Between 2012 and 2013, researchers at University of California, Irvine demonstrated for the first time that BCI technology can restore brain-controlled walking after spinal cord injury. In their study, a person with paraplegia operated a BCI-robotic gait orthosis to regain basic ambulation. In 2009 independent researcher Alex Blainey used the Emotiv EPOC to control a 5 axis robot arm. He made several demonstrations of mind controlled wheelchairs and home automation.

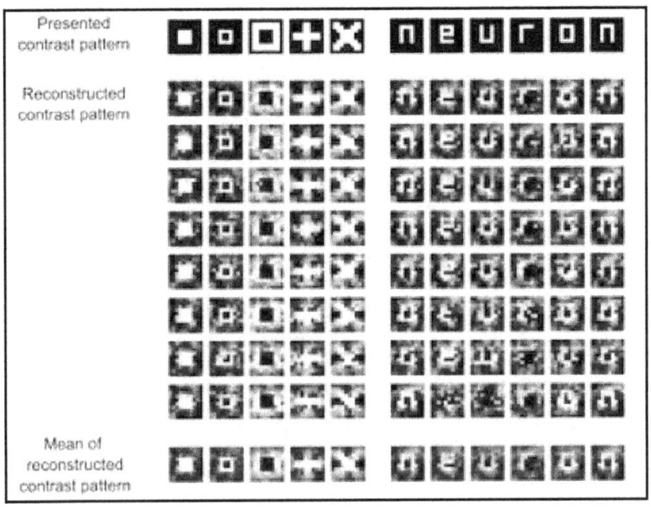

ATR Labs' Reconstruction of Human Vision Using fMRI (Top Row—Original Image; Bottom Row—Reconstruction from Mean of Combined Readings)

Magnetoencephalography and fMRI:

- Magnetoencephalography (MEG) and functional magnetic resonance imaging (fMRI) have both been used as non-invasive BCIs. In a widely reported experiment, fMRI allowed two users to play Pong in real-time by altering their haemodynamic response or brain blood flow through biofeedback.
- fMRI measurements of haemodynamic responses in real time have also been used to control robot arms with a seven-second delay between thought and movement.
- In 2008 research developed in the Advanced Telecommunications Research (ATR) Computational Neuroscience Laboratories in Kyoto, Japan, allowed researchers to reconstruct images from brain signals at a resolution of 10x10 pixels.

- A 2011 study reported second-by-second reconstruction of videos watched by the study's subjects, from fMRI data. This was achieved by creating a statistical model relating videos to brain activity. This model was then used to look up 100 one-second video segments, in a database of 18 million seconds of random YouTube videos, matching visual patterns to brain activity recorded when subjects watched a video. These 100 one-second video extracts were then combined into a mash-up image that resembled the video.

BCI Control Strategies in Neurogaming:

- **Motor Imagery:** Motor imagery involves imagining the movement of body parts, activating the sensorimotor cortex, which modulates sensorimotor oscillations in the EEG. This can be detected by the BCI and used to infer user intent. Motor imagery typically requires training to acquire acceptable control. Training sessions typically consume hours over several days. Regardless of the duration of the training session, users are unable to master the control scheme. This results in very slow pace of the gameplay. Machine learning methods were used to compute a subject-specific model for detecting motor imagery performance. The top performing algorithm from BCI Competition IV in 2022 dataset 2 for motor imagery was the Filter Bank Common Spatial Pattern, developed by Ang et al. from A*STAR, Singapore.
- **Bio/Neurofeedback for Passive BCI Designs:** Biofeedback can be used to monitor a subject's mental relaxation. In some cases, biofeedback does not match EEG, while parameters such as electromyography (EMG), galvanic skin resistance (GSR), and heart rate variability (HRV) can do so. Many biofeedback systems treat disorders such as attention deficit hyperactivity disorder (ADHD), sleep problems in children, teeth grinding, and chronic pain. EEG biofeedback systems typically monitor four brainwave bands (theta: 4–7 Hz, alpha:8–12 Hz, SMR: 12–15 Hz, beta: 15–18 Hz) and challenge the subject to control them. Passive BCI uses BCI to enrich human–machine interaction with information on the user's mental state, for example, simulations that detect when users intend to push brakes during emergency vehicle braking. Game developers using passive BCIs understand that through repetition of game levels the user's cognitive state adapts. During the first play of a given level, the player reacts differently than during subsequent plays: for example, the user is less surprised by an event that they expect.
- **Visual Evoked Potential (VEP):** A VEP is an electrical potential recorded after a subject is presented with a visual stimuli. The types of

VEPs include SSVEPs and P300 potential. Steady-state visually evoked potentials (SSVEPs) use potentials generated by exciting the retina, using visual stimuli modulated at certain frequencies. SSVEP stimuli are often formed from alternating checkerboard patterns and at times use flashing images. The frequency of the phase reversal of the stimulus used can be distinguished by EEG; this makes detection of SSVEP stimuli relatively easy. SSVEP is used within many BCI systems. This is due to several factors. The signal elicited is measurable in as large a population as the transient VEP and blink movement. Electrocardiographic artefacts do not affect the frequencies monitored. The SSVEP signal is robust; the topographic organization of the primary visual cortex is such that a broader area obtains afferents from the visual field's central or fovial region. SSVEP comes with problems. As SSVEPs use flashing stimuli to infer user intent, the user must gaze at one of the flashing or iterating symbols in order to interact with the system. It is, therefore, likely that the symbols become irritating and uncomfortable during longer play sessions. Another type of VEP is the P300 potential. This potential is a positive peak in the EEG that occurs roughly 300 ms after the appearance of a target stimulus (a stimulus for which the user is waiting or seeking) or oddball stimuli. P300 amplitude decreases as the target stimuli and the ignored stimuli grow more similar. P300 is thought to be related to a higher level attention process or an orienting response. Using P300 requires fewer training sessions. The first application to use it was the P300 matrix. Within this system, a subject chooses a letter from a 6 by 6 grid of letters and numbers. The rows and columns of the grid flashed sequentially and every time the selected "choice letter" was illuminated the user's P300 was (potentially) elicited. However, the communication process, at approximately 17 characters per minute, was slow. P300 offers a discrete selection rather than continuous control. The advantage of P300 within games is that the player does not have to learn how to use a new control system, requiring only short training instances to learn gameplay mechanics and the basic BCI paradigm.

Non-Brain-Based Human–Computer interface (Physiological Computing):

- Human-computer interaction can exploit other recording modalities, such as electrooculography and eye-tracking. These modalities do not record brain activity and therefore do not qualify as BCIs.
- **Electrooculography (EOG):** In 1989, a study reported control of a mobile robot by eye movement using electrooculography signals.

A mobile robot was driven to a goal point using five EOG commands, interpreted as forward, backward, left, right, and stop.

- **Pupil-Size Oscillation:** A 2016 article described a new non-EEG-based HCI that required no visual fixation, or ability to move the eyes. The interface is based on covert interest; directing attention to a chosen letter on a virtual keyboard, without the need to look directly at the letter. Each letter has its own (background) circle which micro-oscillates in brightness differently from the others. Letter selection is based on best fit between unintentional pupil-size oscillation and the background circle's brightness oscillation pattern. Accuracy is additionally improved by the user's mental rehearsal of the words 'bright' and 'dark' in synchrony with the brightness transitions of the letter's circle.

Brain-to-Brain Communication

In the 1960s a researcher after training used EEG to create Morse code using alpha waves. On 27 February 2013 Miguel Nicolelis's group at Duke University and IINN-ELS connected the brains of two rats, allowing them to share information, in the first-ever direct brain-to-brain interface. Gerwin Schalk reported that ECoG signals can discriminate vowels and consonants embedded in spoken and imagined words, shedding light on the mechanisms associated with their production and could provide a basis for brain-based communication using imagined speech.

In 2002 Kevin Warwick had an array of 100 electrodes fired into his nervous system in order to link his nervous system to the Internet. Warwick carried out a series of experiments. Electrodes were implanted into his wife's nervous system, allowing them to conduct the first direct electronic communication experiment between the nervous systems of two humans. Other researchers achieved brain-to-brain communication between participants at a distance using non-invasive technology attached to the participants' scalps. The words were encoded in binary streams by the cognitive motor input of the person sending the information. Pseudo-random bits of the information carried encoded words "hola" ("hi" in Spanish) and "ciao" ("goodbye" in Italian) and were transmitted mind-to-mind.

Cell-Culture BCIs

Researchers have built devices to interface with neural cells and entire neural networks in vitro. Experiments on cultured neural tissue focused on building problem-solving networks, constructing basic computers and manipulating robotic devices. Research into techniques for stimulating and recording individual neurons grown on semiconductor chips is neuroelectronics or neurochips. Development of the first neurochip was claimed by a Caltech team led by Jerome Pine and Michael

Maher in 1997. The Caltech chip had room for 16 neurons. In 2003 a team led by Theodore Berger, at the University of Southern California, worked on a neurochip designed to function as an artificial or prosthetic hippocampus.

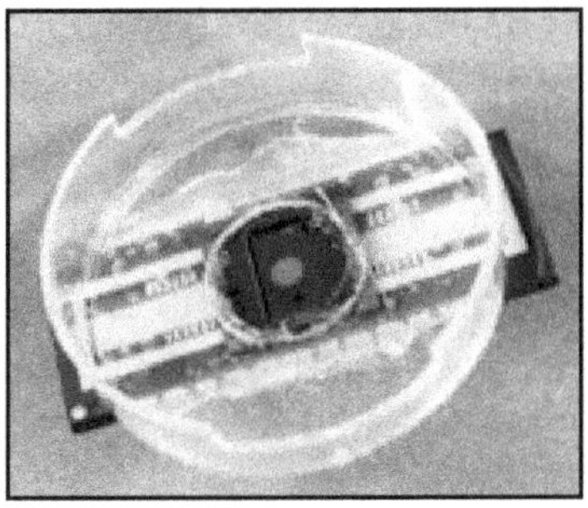

The World's First Neurochip, Developed by Caltech Researchers Jerome Pine and Michael Maher

The neurochip was designed for rat brains. The hippocampus was chosen because it is thought to be the most structured and most studied part of the brain. Its function is to encode experiences for storage as long-term memories elsewhere in the brain. In 2004 Thomas DeMarse at the University of Florida used a culture of 25,000 neurons taken from a rat's brain to fly a F-22 fighter jet aircraft simulator. After collection, the cortical neurons were cultured in a petri dish and reconnected themselves to form a living neural network. The cells were arranged over a grid of 60 electrodes and used to control the pitch and yaw functions of the simulator. The study's focus was on understanding how the human brain performs and learns computational tasks at a cellular level.

Ethical Considerations

Concerns center on the safety and long-term effects on users. These include obtaining informed consent from individuals with communication difficulties, the impact on patients' and families' quality of life, health-related side effects, misuse of therapeutic applications, safety risks, and the non-reversible nature of some BCI-induced changes. Additionally, questions arise about access to maintenance, repair, and spare parts, particularly in the event of a company's bankruptcy.

The legal and social aspects of BCIs complicate mainstream adoption. Concerns include issues of accountability and responsibility, such as claims that BCI influence overrides free will and control over actions, inaccurate translation of cognitive intentions, personality changes resulting from deep-brain stimulation, and the blurring of the line between human and machine. Other concerns involve the use of BCIs in advanced interrogation techniques, unauthorized access ("brain hacking"), social stratification through selective enhancement, privacy issues related to mind-reading, tracking and "tagging" systems, and the potential for mind, movement, and emotion control.

In their current form, most BCIs are more akin to corrective therapies that engage few of such ethical issues. Bioethics is well-equipped to address the challenges posed by BCI technologies, with Clausen suggesting in 2009 that "BCIs pose ethical challenges, but these are conceptually similar to those that bioethicists have addressed for other realms of therapy." Haselager and colleagues highlighted the importance of managing expectations and value.

The evolution of BCIs mirrors that of pharmaceutical science, which began as a means to address impairments and now enhances focus and reduces the need for sleep. As BCIs progress from therapies to enhancements, the BCI community is working to create consensus on ethical guidelines for research, development, and dissemination.

Low-Cost Systems

Various companies are developing inexpensive BCIs for research and entertainment. Toys such as the NeuroSky and Mattel MindFlex have seen some commercial success.

- In 2006, Sony patented a neural interface system allowing radio waves to affect signals in the neural cortex.
- In 2007, NeuroSky released the first affordable consumer based EEG along with the game NeuroBoy. It was the first large scale EEG device to use dry sensor technology.
- In 2008, OCZ Technology developed a device for use in video games relying primarily on electromyography.
- In 2008, Final Fantasy developer Square Enix announced that it was partnering with NeuroSky to create Judecca, a game.
- In 2009, Mattel partnered with NeuroSky to release Mindflex, a game that used an EEG to steer a ball through an obstacle course. It was by far the best selling consumer based EEG at the time.
- In 2009, Uncle Milton Industries partnered with NeuroSky to release the Star Wars Force Trainer, a game designed to create the illusion of possessing the Force.
- In 2009, Emotiv released the EPOC, a 14 channel EEG device that can read 4 mental states, 13 conscious states, facial expressions, and head movements. The EPOC was the first commercial BCI to use dry sensor technology, which can be dampened with a saline solution for a better connection.
- In November 2011, Time magazine selected "necomimi" produced by Neurowear as one of the year's best inventions.

- In February 2014, They Shall Walk (a nonprofit organization fixed on constructing exoskeletons, dubbed LIFESUITs, for paraplegics and quadriplegics) began a partnership with James W. Shakarji on the development of a wireless BCI.
- In 2016, a group of hobbyists developed an open-source BCI board that sends neural signals to the audio jack of a smartphone, dropping the cost of entry-level BCI to £20. Basic diagnostic software is available for Android devices, as well as a text entry app for Unity.
- In 2020, NextMind released a dev kit including an EEG headset with dry electrodes at $399. The device can run various visual-BCI demonstration applications or developers can create their own. It was later acquired by Snap Inc. in 2022.
- In 2023, PiEEG released a shield that allows converting a single-board computer Raspberry Pi to a brain-computer interface for $350.

Future Directions

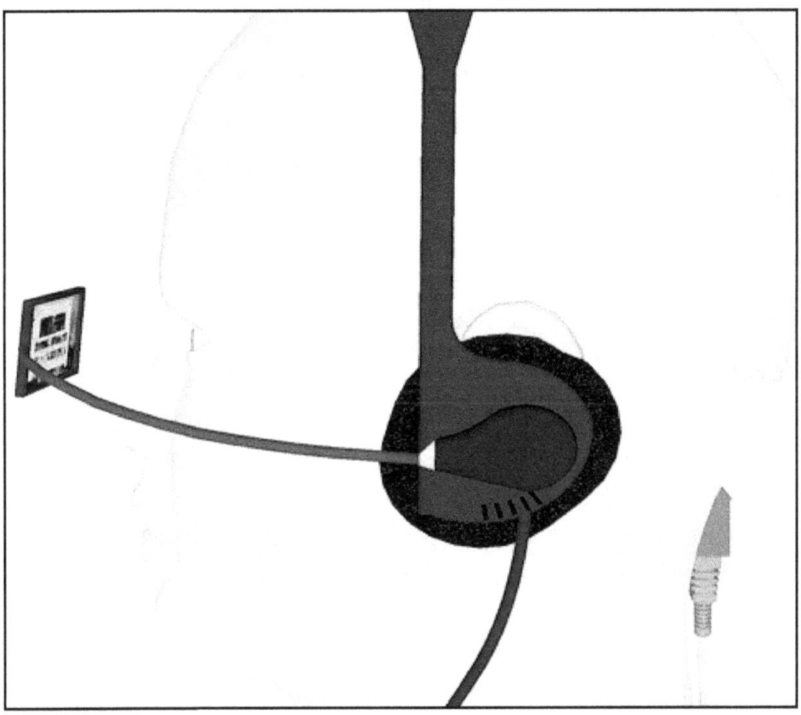

Brain-Computer Interface

A consortium of 12 European partners completed a roadmap to support the European Commission in their funding decisions for the Horizon 2020 framework program. The project was funded by the European Commission. It started in November 2013 and published a roadmap in April 2015. A 2015 publication describes this project, as well as the Brain-Computer Interface Society. It reviewed work within this project that further defined BCIs and applications, explored recent trends, discussed ethical issues, and evaluated directions for new BCIs. Other recent publications too have explored future BCI directions for new groups of disabled users.

Disorders of Consciousness (DOC)

Some people have a disorder of consciousness (DOC). This state is defined to include people in a coma and those in a vegetative state (VS) or minimally conscious state (MCS). BCI research seeks to address DOC. A key initial goal is to identify patients who can perform basic cognitive tasks, which would change their diagnosis, and allow them to make important decisions (such as whether to seek therapy, where to live, and their views on end-of-life decisions regarding them). Patients incorrectly diagnosed may die as a result of end-of-life decisions made by others. The prospect of using BCI to communicate with such patients is a tantalizing prospect.

Many such patients cannot use BCIs based on vision. Hence, tools must rely on auditory and/or vibrotactile stimuli. Patients may wear headphones and/or vibrotactile stimulators placed on responsive body parts. Another challenge is that patients may be able to communicate only at unpredictable intervals. Home devices can allow communications when the patient is ready.

Automated tools can ask questions that patients can easily answer, such as "Is your father named George?" or "Were you born in the USA?" Automated instructions inform patients how to convey yes or no, for example by focusing their attention on stimuli on the right vs. left wrist. This focused attention produces reliable changes in EEG patterns that can help determine whether the patient is able to communicate.

Motor Recovery

People may lose some of their ability to move due to many causes, such as stroke or injury. Research in recent years has demonstrated the utility of EEG-based BCI systems in aiding motor recovery and neurorehabilitation in patients who have had a stroke. Several groups have explored systems and methods for motor recovery that include BCIs. In this approach, a BCI measures motor activity while the patient imagines or attempts movements as directed by a therapist. The BCI may provide two benefits: (1) if the BCI indicates that a patient is not imagining a movement correctly (non-compliance), then the BCI could inform the patient and therapist; and (2) rewarding feedback such as functional stimulation or the movement of a virtual avatar also depends on the patient's correct movement imagery.

So far, BCIs for motor recovery have relied on the EEG to measure the patient's motor imagery. However, studies have also used fMRI to study different changes in the brain as persons undergo BCI-based stroke rehab training. Imaging studies combined with EEG-based BCI systems hold promise for investigating neuroplasticity during motor recovery post-stroke. Future systems might include the fMRI and other measures for real-time control, such as functional near-infrared, probably in tandem with EEGs. Non-invasive brain stimulation has also been explored in combination with BCIs for motor recovery. In 2016, scientists out of the University of Melbourne published preclinical proof-of-concept data related to a potential brain-computer interface technology platform being developed for patients with paralysis to facilitate control of external devices such as robotic limbs, computers and exoskeletons by translating brain activity.

Functional Brain Mapping

In 2014, some 400,000 people underwent brain mapping during neurosurgery. This procedure is often required for people who do not respond to medication. During this procedure, electrodes are placed on the brain to precisely identify the locations of structures and functional areas. Patients may be awake during neurosurgery and asked to perform tasks, such as moving fingers or repeating words. This is necessary so that surgeons can remove the desired tissue while sparing other regions. Removing too much brain tissue can cause permanent damage, while removing too little can mandate additional neurosurgery. Researchers explored ways to improve neurosurgical mapping. This work focuses largely on high gamma activity, which is difficult to detect non-invasively. Results improved methods for identifying key functional areas.

Flexible Devices

Flexible electronics are polymers or other flexible materials (e.g. silk, pentacene, PDMS, Parylene, polyimide) printed with circuitry; the flexibility allows the electronics to bend. The fabrication techniques used to create these devices resembles those used to create integrated circuits and microelectromechanical systems (MEMS). Flexible neural interfaces may minimize brain tissue trauma related to mechanical mismatch between electrode and tissue.

Neural Dust

Neural dust is millimeter-sized devices operated as wirelessly powered nerve sensors that were proposed in a 2011 paper from the University of California, Berkeley Wireless Research Center. In one model, local field potentials could be distinguished from action potential "spikes", which would offer greatly diversified data vs conventional techniques.

CHAPTER 3

Tools and Technologies of Biohacking

3.1 WEARABLES, SENSORS, AND SMART DEVICES

3.1.1 Rise of Wearables in Biohacking

Wearables are among the most visible tools of biohacking, with smartwatches, fitness bands, and health monitors becoming common accessories. These devices initially focused on simple metrics such as step counts and calorie burn, but technological advancements have made them increasingly sophisticated. Today, wearables track heart rate, blood oxygen levels, sleep stages, stress responses, and even electrocardiogram readings. By providing continuous feedback, they allow users to monitor fluctuations in their health and adapt their behaviors accordingly. For biohackers, wearables serve as both motivators and data collectors, encouraging better habits while offering the information needed to experiment with diet, exercise, and sleep optimization. Their integration into everyday routines makes them the cornerstone of modern biohacking practices.

3.1.2 Sensors as Extensions of the Body

While wearables remain the most visible aspect of self-monitoring, sensors provide the deeper technological backbone of biohacking. Sensors can be external, embedded in devices like rings, patches, or bands, or they can be implanted, offering more direct access to biological signals. Continuous glucose monitors are one of the most impactful examples, originally designed for diabetics but now used by biohackers to study how diet and exercise influence blood sugar levels. Similarly, heart rate variability sensors measure subtle changes in the intervals between heartbeats, offering insight into stress resilience and recovery. Other experimental sensors track hydration, ketone levels, and even neurotransmitter activity. By converting biological processes into digital data, sensors extend human self-awareness and provide precise feedback for self-optimization.

3.1.3 Smart Devices and the Connected Ecosystem

Beyond wearables and sensors, smart devices create a broader ecosystem of biohacking tools. Smart scales measure not only body weight but also body fat, muscle mass, and hydration levels, offering a more holistic view of physical health. Smart mattresses and sleep monitors track rest cycles, temperature, and movement, providing insights into sleep quality. Home fitness systems equipped with sensors deliver personalized training feedback, while smart lighting systems adjust circadian rhythms to improve sleep and energy levels. The key strength of these smart devices lies in their ability to integrate with each other, creating interconnected systems where health data flows seamlessly between different platforms. This ecosystem approach reflects the biohacker's vision of a technologically augmented life where the environment itself becomes a tool for self-optimization.

3.1.4 Data and the Quantified Self Movement

Wearables, sensors, and smart devices derive their power not only from their ability to measure but also from their capacity to generate meaningful data. This aligns closely with the quantified self movement, where individuals track everything from sleep hours to blood markers to optimize performance and health. The constant feedback loop created by these devices transforms subjective experiences into objective measurements, allowing for more precise self-experimentation. Biohackers use this data to test hypotheses about their bodies, experimenting with dietary shifts, exercise regimens, or stress management techniques and measuring the outcomes with scientific precision. This culture of self-tracking has expanded beyond enthusiasts into the mainstream, as people increasingly rely on data-driven insights to guide lifestyle decisions.

3.1.5 Applications in Fitness and Performance

Fitness is one of the most immediate areas where wearables and sensors have made an impact. Devices like heart rate monitors, GPS trackers, and smart training systems allow athletes to fine-tune their performance with remarkable accuracy. Biohackers use this information to optimize training zones, prevent overtraining, and accelerate recovery. Metrics such as oxygen saturation and lactate threshold, once accessible only in professional sports labs, are now available to consumers through advanced sensors. Wearables also encourage accountability, motivating users to maintain consistent activity levels. For biohackers who view the body as a system to be optimized, fitness technology serves as a proving ground for experimentation, providing tangible results that reflect lifestyle adjustments.

3.1.6 Applications in Health and Longevity

Beyond fitness, wearables and smart devices are increasingly positioned as tools for preventive health and longevity. Continuous health monitoring allows users to detect early warning signs of illness or chronic conditions. Devices capable of performing electrocardiograms or monitoring oxygen levels can alert users to potential cardiovascular or respiratory issues before they become serious. Sleep trackers reveal patterns that may contribute to fatigue, cognitive decline, or metabolic problems. Longevity-focused biohackers use these insights to adjust diet, exercise, and recovery strategies in pursuit of extending healthspan. In this sense, wearables and sensors serve as personal health companions, making preventive care an everyday activity rather than a reactive measure.

3.1.7 Psychological and Behavioral Dimensions

The influence of wearables and smart devices extends beyond biological measurements into psychological and behavioral realms. The very act of tracking creates heightened awareness, often motivating individuals to make healthier choices. Gamification features, such as badges, streaks, and progress goals, add a layer of motivation and accountability. Social sharing options allow users to compare progress with peers, fostering community engagement and support. However, this psychological dimension also raises concerns, as some users may experience anxiety, obsession, or dependency on devices. Biohackers often grapple with finding balance, ensuring that technology enhances rather than dominates their relationship with their bodies. The psychological impact of constant tracking remains one of the most complex aspects of wearable adoption.

3.1.8 Privacy, Security, and Ethical Challenges

As wearables and sensors collect increasingly sensitive health data, privacy and security issues become unavoidable. Data stored on devices or shared through apps may be vulnerable to hacking or misuse. Companies often aggregate and sell anonymized health data, raising ethical concerns about consent and ownership. For biohackers, who embrace transparency and self-experimentation, the trade-off between data access and privacy can be a difficult one. Ethical challenges also arise in medical contexts, where insurers or employers could potentially use wearable data to make decisions about coverage or employment. Addressing these issues requires robust regulatory frameworks and ethical guidelines that balance innovation with individual rights.

3.2 MOBILE APPS FOR QUANTIFIED SELF MONITORING

The philosophy of the Quantified Self—the pursuit of self-knowledge through numbers—has found its most potent and pervasive expression in the smartphone. This ubiquitous device, carried in billions of pockets worldwide, has become the central nervous system for the modern biohacker, and the mobile applications it hosts are the tools that make rigorous self-tracking not just possible, but practical. These apps transform the abstract concept of self-optimization into a daily, actionable practice, providing a digital dashboard that renders the complexities of human biology into streams of intelligible data. They are the indispensable bridge between the biohacker's ambition and their lived reality, enabling a continuous loop of measurement, analysis, and intervention that was once the exclusive domain of research laboratories and elite athletes. The rise of these applications represents the mass democratization of biohacking, placing an unprecedented depth of personal analytics into the hands of anyone with a mobile phone.

3.2.1 Ecosystem of Tracking: Categories of Self-Monitoring Apps

The landscape of mobile apps for Quantified Self monitoring is vast and diverse, catering to every conceivable aspect of human physiology and behavior. This ecosystem can be broadly categorized by the type of data they capture and the domains they seek to optimize. Foundational to this are health and wellness apps that sync with wearable devices, aggregating metrics such as step count, heart rate, sleep stages, and blood oxygen saturation. Platforms like Apple Health and Google Fit act as central repositories, corralling data from various sources to provide a holistic view of daily activity and rest.

A second major category encompasses nutrition and metabolic tracking. Apps like MyFitnessPal and Cronometer allow users to log every consumed morsel, providing detailed breakdowns of macronutrient and micronutrient intake, and often integrating with glucose monitors for real-time metabolic feedback. A third, rapidly growing category is focused on cognitive and emotional states. Meditation apps such as Headspace and Calm track session duration and consistency, while mood journaling apps like Daylio enable users to identify correlations between their activities, sleep, and emotional well-being. Together, these applications create a multi-dimensional portrait of the user, quantifying everything from physical exertion and nutritional input to mental calm and emotional resilience.

3.2.2 Technology Enabling the Quantification: Sensors and Synergy

The power of these mobile applications is not inherent to the software alone; it is derived from their symbiotic relationship with a growing array of hardware sensors and devices. The smartphone itself is a sophisticated data-gathering tool, equipped with accelerometers to measure movement, microphones for audio analysis, and location services to track distance and pace. However, its true potential is unlocked when paired with external wearables.

Smartwatches and fitness bands act as perpetual data collection nodes on the body, providing continuous, passive biometric monitoring that would be impossible through manual input alone. This synergy creates a powerful feedback loop: the wearable collects raw physiological data, the mobile app processes and visualizes this information, and the user derives insight to inform their choices. Furthermore, the ecosystem is expanding to include more specialized medical-grade devices that connect via Bluetooth. Smart scales measure body composition, continuous glucose monitors (CGMs) stream blood sugar levels, and wearable ECG patches record heart rhythm. The mobile app serves as the central command hub for this entire personal biometry network, integrating disparate data streams to reveal patterns and correlations that would remain invisible if each metric were viewed in isolation.

3.2.3 From Data to Insight: The Analytical Power of Modern Applications

The evolution of these tools has moved far beyond simple data logging. The most advanced applications now incorporate sophisticated algorithms and machine learning to transform raw data into actionable intelligence. They no longer just report that you took 10,000 steps; they analyze the intensity and pattern of your activity over time to assess your cardiovascular fitness level (VO2 Max estimate). They don't just record your sleep; they compare your sleep duration and quality against population norms and provide a personalized sleep score with tailored recommendations for improvement.

This analytical layer is the core of their value as biohacking tools. Nutrition apps can predict the glycemic load of a meal before it is eaten, while mood tracking apps can generate reports showing that your anxiety levels are statistically lower on days when you exercise for more than thirty minutes. By identifying these correlations, the apps move the user from passive observation to active experimentation. They empower the individual to form hypotheses—"If I stop drinking caffeine after 2 PM, will my deep sleep increase?"—and use their own data to find the answer, embodying the scientific method in the pursuit of personal well-being.

3.2.4 Challenges and Ethical Considerations of a Quantified Life

Despite their transformative potential, mobile apps for self-quantification are not without significant challenges and ethical dilemmas. The most pressing concern is data privacy and security. These applications collect the most intimate details of a person's life—their health vulnerabilities, daily routines, and location history. Questions abound regarding who owns this data, how it is stored, and how it might be used by the app developers, third-party advertisers, or even insurers and employers. The potential for data breaches or the use of this information for discriminatory practices is a serious risk.

Another major concern is the psychological impact of constant self-monitoring. The pursuit of optimization can easily spiral into orthorexia or obsessive-compulsive behaviors, where life becomes a relentless pursuit of perfect metrics, leading to anxiety and a disconnection from intuitive bodily senses. The quantified life can create a paradox where one is so focused on measuring the experience of living that one forgets to actually live it. Furthermore, the accuracy of these consumer-grade tools is often questionable, potentially leading users to make important health decisions based on unreliable data.

3.3 DATA PLATFORMS FOR SELF-EXPERIMENTATION

The biohacking movement has evolved from a niche subculture into a mainstream approach to personal optimization, driven by the growing availability of data platforms and self-experimentation tools. Unlike traditional medical research, which relies on controlled clinical trials, biohackers take a hands-on approach, using their own bodies as laboratories to test dietary interventions, supplement regimens, sleep protocols, and even genetic modifications.

This shift has been made possible by the proliferation of wearable devices, mobile applications, and open-source data platforms that allow individuals to track, analyze, and optimize their biological metrics in real time. The democratization of these tools has empowered people to move beyond anecdotal health advice and instead rely on quantifiable, personalized data to guide their wellness journeys. As a result, self-experimentation has become a cornerstone of the biohacking ethos, blending scientific rigor with individual curiosity to unlock new levels of human performance and well-being.

3.3.1 Wearable Devices: The Foundation of Biohacking Data

Wearable technology has become the backbone of self-experimentation, providing continuous, real-time data on physiological and behavioral metrics. Devices such as the Oura Ring, Whoop Strap, and Apple Watch have revolutionized how individuals monitor their health, offering insights into heart rate variability, sleep quality, activity levels, and even blood oxygen saturation. These wearables go beyond simple step counting, using advanced sensors and algorithms to detect patterns in stress, recovery, and metabolic function.

For biohackers, this data is invaluable, as it allows for precise adjustments to lifestyle factors such as exercise timing, meal composition, and sleep schedules. The integration of artificial intelligence into these platforms further enhances their utility, enabling predictive analytics that can forecast energy levels, recovery needs, and even illness susceptibility. As wearable technology continues to advance, with the development of non-invasive glucose monitors and advanced EEG headbands, the depth and accuracy of self-experimentation data will only increase, providing biohackers with unprecedented insights into their biological systems.

3.3.2 Mobile Applications and Digital Platforms for Self-Tracking

Beyond wearables, mobile applications and digital platforms have become essential tools for biohackers engaged in self-experimentation. Apps like Cronometer and MyFitnessPal allow users to meticulously track macronutrient and micronutrient intake, ensuring that dietary interventions are precisely measured and adjusted. Sleep tracking apps such as Sleep Cycle and Pillow provide detailed analyses of sleep stages, offering recommendations for optimizing rest and recovery. For those exploring intermittent fasting or time-restricted eating, apps like Zero and Life Fasting Tracker help monitor fasting windows and their effects on energy, cognition, and metabolic health. Additionally, platforms like InsideTracker and Nutrisense offer advanced blood biomarker analysis, enabling users to track changes in cholesterol, inflammation markers, and other critical health indicators over time. These digital tools transform raw biological data into actionable insights, allowing biohackers to refine their protocols based on empirical evidence rather than guesswork.

3.3.3 Open-Source Data Platforms and Citizen Science

The open-source movement has played a pivotal role in the evolution of self-experimentation, providing biohackers with platforms to share data, collaborate on projects, and contribute to collective knowledge. Websites like Open Humans and the Quantified Self community allow individuals to upload their biological data, participate in citizen science projects, and compare their results with others engaged in similar experiments. These platforms foster a culture of transparency and collaboration, where users can access shared datasets, replicate experiments, and refine their own protocols based on aggregated insights.

The open-source ethos extends to software as well, with tools like Python-based data analysis scripts and R programming environments enabling biohackers to customize their data interpretation and visualization. This democratization of data not only accelerates personal optimization but also contributes to broader scientific understanding, as large-scale self-reported datasets can reveal trends and correlations that might be overlooked in traditional research settings.

3.3.4 Advanced Biometric Monitoring and Lab Testing

For biohackers seeking deeper biological insights, advanced biometric monitoring and lab testing have become indispensable tools. Continuous glucose monitors (CGMs) like those from Nutrisense and Levels provide real-time feedback on how different foods and activities affect blood sugar levels, offering a window into metabolic health that was once only available in clinical settings. Similarly, wearable EEG devices such as Muse and NeuroSky allow users to track brainwave activity, enabling experiments with meditation, nootropic supplements, and cognitive training protocols.

Beyond wearables, direct-to-consumer lab testing services have expanded the scope of self-experimentation, with companies like InsideTracker and WellnessFX offering comprehensive blood panels that measure everything from hormone levels to micronutrient status. These tests provide a baseline for health metrics, allowing biohackers to track longitudinal changes and adjust their interventions with precision. The integration of these advanced monitoring tools with wearable and app-based data creates a holistic picture of biological function, empowering individuals to make informed, data-driven decisions about their health.

3.3.5 Role of Artificial Intelligence in Self-Experimentation

Artificial intelligence is increasingly becoming a game-changer in the realm of biohacking, offering predictive analytics and personalized recommendations based on vast datasets. AI-driven platforms can analyze patterns in sleep, activity, diet, and biomarkers, identifying correlations and suggesting optimizations that might not be apparent through manual analysis. For example, AI algorithms can predict how specific dietary changes might affect energy levels or sleep quality based on historical data, allowing for proactive adjustments.

Similarly, machine learning models can identify optimal workout times, recovery protocols, and even supplement dosages tailored to an individual's unique physiology. The integration of AI into self-experimentation tools is making biohacking more accessible and effective, as these systems can process complex datasets and provide actionable insights without requiring users to have advanced statistical knowledge. As AI continues to evolve, its role in biohacking will likely expand, enabling even more precise and personalized health optimization strategies.

3.4 LABORATORY TOOLS FOR DIY BIOTECH

3.4.1 Democratization of Biotechnology

The availability of affordable laboratory equipment has shifted biotechnology from an exclusive domain into a more open field. Once prohibitively expensive, tools such as centrifuges, microscopes, and PCR machines are now accessible in portable, low-cost forms. This democratization reflects the broader ethos of the biohacking community, which values open-source knowledge, collaborative exploration, and hands-on experimentation. Makerspaces and community laboratories have sprung up in cities across the globe, providing access to equipment and training for those who wish to explore biotechnology outside traditional academic or corporate structures. By lowering the barriers of entry, these tools make it possible for ordinary individuals to participate in scientific discovery and bio-innovation.

3.4.2 Microscopes and the Window into Life

One of the most fundamental tools in any biotechnology lab is the microscope. For DIY biotechnologists, microscopes serve as an essential gateway into the cellular and microbial world. Affordable digital microscopes allow individuals to observe bacteria, yeast, or human cells, providing both educational and experimental insights. Microscopy enables DIY biologists to study morphology, track cellular changes, and even verify experimental results. Advanced hobbyists sometimes acquire more powerful models, such as fluorescence microscopes, which reveal molecular structures through dyes and light. The microscope embodies the spirit of DIY biotech by translating abstract biological concepts into visible, tangible realities that anyone can explore.

3.4.3 PCR Machines and DNA Amplification

Polymerase chain reaction, or PCR, is one of the most transformative tools in biotechnology, enabling the amplification of DNA sequences for analysis and experimentation. For decades, PCR was confined to professional laboratories due to the cost and complexity of thermal cyclers. Today, however, compact and affordable PCR machines are available to DIY biologists. These devices allow biohackers to extract, replicate, and study DNA samples from various organisms. Applications range from identifying genetic markers in plants to exploring bacterial DNA. PCR is also the foundation for gene editing projects, making it one of the central tools in DIY biotech. The ability to work directly with DNA marks a significant leap for hobbyists, turning abstract genetic information into a manipulable material for experimentation.

3.4.4 Gel Electrophoresis and DNA Visualization

Once DNA is amplified, it must be analyzed, and gel electrophoresis provides a simple yet powerful method for visualizing genetic material. In this technique, DNA fragments are separated based on size as they move through a gel under the influence of an electric field. For DIY biologists, gel electrophoresis kits are widely available and relatively inexpensive, making them a staple of amateur biotechnology labs. This tool allows hobbyists to verify PCR results, identify specific genetic fragments, and gain hands-on experience with the molecular foundations of biology. While professional labs use advanced versions with sophisticated imaging systems, DIY kits provide accessible entry points into genetic analysis and experimentation.

3.4.5 Centrifuges and Cellular Separation

Centrifuges are another indispensable tool, used to separate biological materials based on density. By spinning samples at high speeds, centrifuges allow biohackers to isolate cellular components, such as DNA, proteins, or organelles. Small-scale centrifuges, often marketed to educational institutions, are affordable and safe for community labs or home experiments. Some DIY enthusiasts have even designed open-source centrifuge models that can be assembled with 3D printers and inexpensive parts. The centrifuge expands the capabilities of DIY biologists, enabling more complex experiments that require separation of cellular materials for downstream analysis.

3.4.6 CRISPR Kits and At-Home Gene Editing

Perhaps the most striking development in DIY biotechnology is the availability of

CRISPR gene editing kits. CRISPR-Cas9 technology has revolutionized biology by providing a relatively simple method for cutting and modifying DNA sequences. Today, educational CRISPR kits allow hobbyists to experiment with gene editing in harmless organisms such as bacteria or yeast. These kits provide instructions, reagents, and tools necessary for simple gene-editing experiments, making one of the most advanced technologies in modern science accessible to non-specialists. While these kits are designed for safety and limited scope, they symbolize the broader potential of DIY biotech to bring cutting-edge science into everyday spaces.

3.4.7 Incubators and Culturing Organisms

In addition to molecular tools, DIY biotechnology often involves working with living organisms such as bacteria, yeast, or algae. Incubators provide controlled environments for growing and maintaining these organisms under specific conditions. Compact, low-cost incubators are widely available, and some community labs design open-source versions that can be built with basic materials. Culturing microorganisms allows biohackers to experiment with fermentation, synthetic biology, or environmental sampling. Incubators expand the range of possible projects, from testing antimicrobial properties of substances to exploring metabolic pathways in microbes.

3.4.8 Biosensors and Analytical Tools

Another growing category of DIY biotech tools includes biosensors and portable analytical devices. These tools allow hobbyists to measure biological markers such as pH, glucose, or oxygen levels in real time. Portable spectrophotometers, for example, enable analysis of protein concentration or microbial growth by measuring light absorption. Such tools enhance experimental precision and provide a bridge between amateur exploration and professional-level data collection. In the hands of DIY biohackers, biosensors are used not only for experiments but also for personal biohacking applications, such as monitoring metabolic responses to diet or supplements.

3.4.9 Ethical and Safety Considerations

The proliferation of laboratory tools for DIY biotech raises important ethical and safety questions. While most DIY biologists work with safe organisms and strictly follow biosafety guidelines, the potential for misuse or accidents cannot be ignored. Community labs often emphasize responsible practices, providing training in biosafety and creating environments where experimentation can occur under supervision. Critics worry, however, that the availability of gene editing kits or microbial culture equipment could lead to unintended risks if misused. Ethical debates also focus on the blurring of lines between professional science and amateur exploration, raising questions about responsibility, regulation, and oversight. The success of DIY biotech depends on balancing accessibility with safeguards that protect both individuals and the public.

3.4.10 Role of Community and Collaboration

A defining feature of DIY biotech is its communal nature. While some enthusiasts set up personal laboratories, most participate in community spaces where equipment, training, and expertise are shared. These labs foster collaboration, creativity, and education, often functioning as hubs where biology enthusiasts from diverse backgrounds can meet. Collaborative projects, workshops, and citizen science initiatives emerge from these spaces, highlighting the collective spirit of the DIY biohacking movement. Community labs not only provide access to expensive equipment but also promote a culture of responsibility and knowledge sharing, which is crucial for sustaining safe and meaningful DIY biotechnology.

3.5 NEUROSTIMULATION AND BIOFEEDBACK DEVICES

3.5.1 Principles of Neural Modulation: How the Tools Work

Neurostimulation and biofeedback devices, though distinct in function, share a common goal: to alter brain state and function for a desired outcome. Neurostimulation techniques work on the principle of entrainment or neuromodulation. Technologies like transcranial Direct Current Stimulation (tDCS) apply a low, constant electrical current to specific regions of the scalp via electrodes. This current does not cause neurons to fire but rather modulates their likelihood of firing, potentially enhancing cortical excitability in areas responsible for learning or attention for a period following stimulation.

Conversely, cranial electrotherapy stimulation (CES) uses a microcurrent across the head, often from ear to ear, which is believed to influence neurotransmitter production, particularly those related to mood and anxiety like serotonin and dopamine. Biofeedback, on the other hand, operates on a principle of mirroring. Electroencephalogram (EEG) biofeedback, or neurofeedback, involves placing sensors on the scalp to monitor brainwave patterns—such as beta waves (associated with focus), alpha waves (calm alertness), and theta waves (drowsiness or meditation). This information is fed to the user in real-time through a visual or auditory display on a computer or mobile device. Through this feedback loop, the user learns, through operant conditioning, to consciously shift their brainwave activity toward a desired state, effectively training their brain as one would train a muscle.

3.5.2 Toolbox for the Mind: Key Devices and Technologies

The market for consumer-grade neural devices has expanded rapidly, offering a range of tools for various budgets and goals. In the realm of neurostimulation, tDCS devices like the Halo Sport (now Halo Neuroscience) gained popularity among athletes and learners by aiming to prime the motor cortex for skill acquisition.

More recently, transcranial Alternating Current Stimulation (tACS) devices have emerged, which use rhythmic electrical currents to potentially synchronize brainwave patterns, such as boosting alpha waves for relaxation or gamma waves for cognitive processing. For biofeedback, the landscape is dominated by wearable EEG headsets. Consumer-friendly devices like Muse and NeuroSky provide a simplified form of neurofeedback, often packaged within meditation apps that give auditory feedback—such as calming weather sounds during calm states—to guide the user into a meditative state.

More advanced clinical-grade systems used by practitioners offer higher resolution and greater control for addressing specific conditions. Additionally, heart rate variability (HRV) biofeedback devices, while not measuring the brain directly, are a crucial part of this ecosystem. By teaching users to regulate their breathing to maximize HRV—a key marker of autonomic nervous system balance—they indirectly promote a mental state of calm focus and resilience, demonstrating the intricate connection between cardiovascular and cognitive function.

3.5.3 Applications and Pursued Benefits: From Peak Performance to Mental Wellness

The applications for these technologies are as diverse as the biohacking community itself. A primary use case is the enhancement of cognitive performance. Biohackers employ tDCS to potentially accelerate learning curves for complex skills like language acquisition or musical instrument practice, while neurofeedback sessions are used to train sustained focus, reducing susceptibility to distraction in an increasingly interrupt-driven world. For athletes and executives, these tools offer a legal and non-pharmaceutical edge in achieving a state of "flow" or peak performance on demand.

Equally significant is the application for mental wellness and emotional regulation. CES devices are marketed for reducing symptoms of anxiety, depression, and insomnia by promoting a calmer neurological baseline. Neurofeedback has shown promise in helping individuals manage ADHD, PTSD, and chronic stress by teaching the brain to downregulate hyper-aroused states and upregulate calm, focused ones. Furthermore, these devices are used to optimize recovery and sleep. Certain neurostimulation protocols are designed to promote slow-wave sleep, the most restorative phase, while biofeedback training can help quiet the mind before bed, easing the transition into sleep and improving overall sleep architecture.

3.5.4 Critical Considerations and Responsible Use

The burgeoning field of consumer neurotechnology is not without serious questions and potential risks. The foremost concern is the lack of robust, long-term scientific research for many consumer devices. While the underlying principles are based on solid neuroscience, the specific protocols, dosage, and long-term effects of daily use in healthy individuals are often not well-established. The brain is an exceptionally complex and delicate system, and applying electrical currents or training specific patterns without professional guidance could potentially lead to unintended consequences, such as mood changes, increased anxiety, or disrupted sleep patterns.

The "do-it-yourself" nature of the biohacking ethos can sometimes conflict with the need for caution when manipulating brain function. Furthermore, the placebo effect is a powerful confounding variable; a user's belief in the technology's efficacy can itself produce positive results, making it difficult to discern the true effect of the device. Ethical considerations also arise regarding cognitive enhancement and the potential for a new form of inequality—a "neuro-divide"—between those who can afford these tools and those who cannot.

CHAPTER 4

Cognitive and Performance Enhancement

4.1 SMART DRUGS (NOOTROPICS): NATURAL AND SYNTHETIC

4.1.1 Origins of Nootropics

The term nootropic was coined in the 1970s by Romanian chemist Corneliu Giurgea, who described these compounds as substances that enhance learning and memory while being safe and non-toxic. However, the use of cognitive enhancers stretches back centuries, as humans have long experimented with plants and foods believed to improve mental clarity. Caffeine, one of the most widely used nootropics, has been consumed in tea and coffee for centuries. Other traditional substances include ginseng, bacopa monnieri, and gotu kola, which were used in traditional Chinese and Ayurvedic medicine to sharpen the mind and improve memory. The formalization of nootropics as a field, however, emerged with the development of synthetic compounds like piracetam, which marked the beginning of modern smart drug research.

4.1.2 Natural Nootropics and their Effects

Natural nootropics draw their appeal from their organic origins and long histories of traditional use. One of the most widely consumed natural nootropics is caffeine, a stimulant that enhances alertness and reduces fatigue by blocking adenosine receptors in the brain. L-theanine, found in green tea, is often combined with caffeine to promote calm focus without jitteriness. Herbal supplements such as ginkgo biloba are believed to improve blood circulation in the brain and enhance memory, while bacopa monnieri has been studied for its ability to reduce anxiety and support learning. Adaptogens such as rhodiola rosea help the body manage stress, indirectly supporting cognitive performance. These natural compounds are often marketed as safe, holistic, and gentle alternatives to pharmaceuticals, making them attractive to biohackers and health-conscious individuals.

4.1.3 Synthetic Nootropics and their Development

Synthetic nootropics, on the other hand, are laboratory-designed compounds specifically engineered to enhance cognition. The first of these was piracetam, developed in the 1960s, which became the foundation for an entire class of racetams. Racetams, such as aniracetam and oxiracetam, are thought to modulate neurotransmitter activity, improving memory and focus. Another class of synthetic nootropics includes prescription stimulants such as modafinil, originally developed for sleep disorders like narcolepsy but now widely used off-label to boost alertness and productivity. Adderall and Ritalin, prescribed for attention deficit hyperactivity disorder, are also sometimes misused as cognitive enhancers. These synthetic compounds often produce stronger effects than natural nootropics, but they also carry higher risks of side effects, dependency, and ethical controversy.

4.1.4 Mechanisms of Action

Both natural and synthetic nootropics work by influencing brain chemistry, though their mechanisms vary widely. Some nootropics enhance neurotransmitter activity, particularly acetylcholine, dopamine, and serotonin, which play roles in learning, motivation, and mood. Others increase cerebral blood flow, ensuring that the brain receives adequate oxygen and nutrients to function optimally. Certain nootropics act as antioxidants, protecting brain cells from oxidative stress and age-related decline. Stimulant-based nootropics, such as caffeine or modafinil, promote wakefulness by modulating sleep-wake neurotransmitters. Adaptogens, meanwhile, influence stress hormones like cortisol, indirectly supporting mental clarity. The diversity of mechanisms reflects the complexity of the brain and the many pathways through which cognitive performance can be enhanced.

4.1.5 Applications of Nootropics in Daily Life

The appeal of nootropics lies in their wide-ranging applications for both healthy individuals and those managing cognitive challenges. Students often use them to improve focus and memory during periods of intense study, while professionals turn to them for sustained productivity and creativity. Entrepreneurs and innovators may experiment with nootropics to maintain mental stamina in high-pressure environments. Some biohackers incorporate nootropics into broader regimens that include sleep optimization, diet, and exercise, seeking holistic cognitive enhancement. On the medical side, nootropics are studied for their potential to treat neurodegenerative diseases like Alzheimer's, where they may slow cognitive decline. This dual role as both lifestyle enhancers and therapeutic agents contributes to the growing popularity of smart drugs.

4.1.6 Risks and Limitations of Nootropics

Despite their appeal, both natural and synthetic nootropics come with limitations and risks. Natural nootropics are often perceived as safe, but their effectiveness can vary, and some herbal supplements may interact with medications or cause side effects. Synthetic nootropics can offer stronger effects but often pose greater risks. Stimulant-based drugs may lead to insomnia, anxiety, cardiovascular strain, or dependency if misused. Even racetams, though generally considered low in toxicity, lack comprehensive long-term safety data. The placebo effect also complicates evaluation, as many users report subjective improvements that may not correspond to measurable changes. These limitations highlight the importance of caution, informed experimentation, and medical guidance when exploring nootropics.

4.1.7 Ethical and Societal Considerations

The use of smart drugs raises significant ethical questions about fairness, accessibility, and human identity. In academic and professional settings, the use of prescription stimulants for performance enhancement blurs the line between therapy and unfair advantage. Critics argue that widespread use of cognitive enhancers could create pressure to conform, with individuals feeling compelled to take nootropics to remain competitive. Issues of accessibility also arise, as wealthier individuals may have greater access to expensive synthetic nootropics, potentially widening social inequalities. On a broader philosophical level, nootropics challenge traditional notions of human effort and achievement, prompting debates about whether enhanced cognition undermines authenticity or represents legitimate progress.

4.1.8 Role of Biohacking and Self-Experimentation

Within the biohacking community, nootropics occupy a central role as tools for self-experimentation. Biohackers often combine different compounds into "stacks" designed to target multiple aspects of cognition simultaneously. For example, a common stack combines caffeine for alertness with L-theanine for calm focus, and racetams for memory enhancement. Online communities share experiences, protocols, and research findings, creating collective knowledge around nootropic experimentation. While this culture fosters innovation, it also raises concerns about unregulated use and reliance on anecdotal evidence. The experimental ethos of biohackers highlights both the potential and the uncertainties of nootropics, illustrating how personal experimentation drives the field forward.

4.2 ENHANCING FOCUS, MEMORY, AND CREATIVITY

The pursuit of peak mental performance is a central pillar of the biohacking movement, driven by a desire to not only correct cognitive deficits but to amplify innate abilities to their fullest potential. This endeavor focuses on the interconnected triad of focus, memory, and creativity—three distinct yet deeply intertwined facets of human cognition. Focus, or sustained attention, is the gateway through which information enters our conscious awareness. Memory is the mechanism by which we encode, store, and retrieve that information, forming the foundation of knowledge and experience. Creativity is the highest-order function, the ability to synthesize stored information in novel and valuable ways. Enhancing this cognitive trinity is not about simply working harder; it is about working smarter by understanding and optimizing the biological and environmental systems that govern mental performance. It is a holistic approach that recognizes the brain not as a static organ, but as a dynamic, plastic system that can be trained, nourished, and fine-tuned.

4.2.1 Neurobiological Foundation: How the Brain Supports Cognition

To effectively enhance cognitive function, one must first understand its biological underpinnings. Focus is largely governed by the prefrontal cortex, the brain's executive control center, and is heavily influenced by neurotransmitter systems. Dopamine and norepinephrine play crucial roles in maintaining alertness, motivation, and the ability to filter out distractions. The cholinergic system, reliant on acetylcholine, is fundamental for learning and memory formation, particularly in the hippocampus, the brain's primary memory encoding structure.

Creativity, however, is a more distributed process. It involves the default mode network (DMN), which is active during mind-wandering and spontaneous thought, and the executive control network, which provides focused direction. The interplay between these seemingly opposed networks—the free-flowing association of the DMN and the disciplined focus of the executive network—is where novel ideas are believed to emerge. Furthermore, all cognitive functions depend on fundamental biological support: stable energy supply from glucose metabolism, efficient communication between neurons across synapses, and the reduction of inflammation that can impair neural efficiency.

4.2.2 Strategic Interventions for Laser-Sharp Focus

In an age of relentless distraction, cultivating deep focus is a superpower. Biohacking concentration involves both removing impediments and actively enhancing the brain's ability to sustain attention. A primary strategy is the management of the neurotransmitter dopamine, not by flooding the system, but by optimizing its receptors and preventing burnout. This involves strategies like "dopamine fasting" or scheduling breaks from highly stimulating activities to reset the brain's reward circuitry and reduce the craving for novelty.

From a nutritional standpoint, compounds that support catecholamine production are key. L-Tyrosine, a precursor to dopamine and norepinephrine, has been shown to improve cognitive flexibility and performance under stress. Similarly, natural stimulants like caffeine, when used strategically and paired with L-Theanine from green tea, can enhance alertness without the accompanying jitters and anxiety. Beyond chemistry, environmental design is critical. The use of noise-cancelling headphones, dedicated deep work blocks calendared like important meetings, and the application of the Pomodoro Technique (25 minutes of focused work followed by a 5-minute break) are all behavioral biohacks that reduce cognitive load and train the muscle of attention, minimizing the energy required to maintain focus.

4.2.3 Fortifying the Archives: Techniques for Robust Memory

Memory is a process, not a static recording, and each stage—encoding, consolidation, and retrieval—can be optimized. Encoding, the initial learning of information, is enhanced by deep, attentive processing. Techniques like spaced repetition, which involves reviewing information at increasing intervals, actively combat the forgetting curve and are powerfully facilitated by software like Anki.

Elaborative encoding, which involves connecting new information to existing knowledge networks, creates more neural hooks for later retrieval. The consolidation of memory, where short-term memories are stabilized into long-term storage, occurs primarily during sleep, specifically during deep slow-wave sleep and REM sleep. Therefore, the single most effective biohack for memory is prioritizing sleep quality and duration.

Nutritionally, the cholinergic system can be directly supported. Choline precursors like Alpha-GPC and Citicoline have been shown to support memory formation and recall, particularly in combination with other nootropics. Furthermore, compounds that support Brain-Derived Neurotrophic Factor (BDNF), such as Lion's Mane Mushroom and the flavonoid apigenin, act like fertilizer for the brain, promoting the growth of new neurons and synapses in the hippocampus, directly enhancing the brain's structural capacity for memory.

4.2.4 Cultivating the Unpredictable: Fostering Creativity

Enhancing creativity is a more paradoxical challenge than improving focus or memory, as it requires fostering both disciplined focus and spontaneous, unfettered association. The biohacker's approach involves creating the optimal internal and external conditions for creative insight to emerge, rather than trying to force it. The first step is often saturation—immersing oneself deeply in the problem or domain, loading the conscious and unconscious mind with relevant information.

This is followed by a crucial incubation period, where conscious effort is ceased. Engaging in activities that relax the prefrontal cortex and activate the DMN—such as walking in nature, taking a warm shower, or practicing meditation—allows the unconscious mind to make novel connections between disparate ideas. Nutritionally, some evidence suggests that compounds that mildly dampen prefrontal control and encourage mental flexibility, such as the microdosing of psychedelics (though legally fraught and under research) or even theanine, can promote a state of "hypofrontality" conducive to creative insight. Ultimately, the best biohack for creativity is to recognize and respect its rhythm: alternating intense periods of focused work with deliberate periods of restorative mind-wandering and play, creating the mental space for the "Aha!" moment to strike.

4.2.5 Synergistic System: Integration and Lifestyle

Ultimately, focus, memory, and creativity do not exist in isolation; they are part of a synergistic cognitive system. The most powerful biohacking protocol is one that supports the entire foundation of brain health. This includes consistent aerobic exercise, which boosts BDNF, increases cerebral blood flow, and reduces stress hormones. It involves a nutrient-dense diet rich in antioxidants and anti-inflammatory fats to protect neural tissue from oxidative stress.

It requires effective stress management through practices like mindfulness and meditation, which thicken the prefrontal cortex and improve emotional regulation, preventing amygdala hijackings that shut down higher-order thinking. It is this holistic integration—the combination of targeted nootropics, strategic behavioral techniques, optimal environmental design, and foundational physical health—that creates the conditions for the cognitive trinity to truly flourish. By viewing the brain as a garden to be cultivated rather than a machine to be exploited, biohackers can achieve a state of sustainable cognitive flow, enhancing not just productivity, but the very quality of thought and experience itself.

4.3 NEUROPLASTICITY AND BRAIN TRAINING

The human brain, once thought to be a fixed and unchangeable organ after early development, is now recognized as a dynamic and malleable system capable of remarkable transformation throughout life. This capacity for change, known as neuroplasticity, refers to the brain's ability to reorganize itself by forming new neural connections, strengthening existing pathways, and even generating new neurons in response to experience, learning, and environmental demands.

Neuroplasticity challenges the long-held belief that brain development is confined to childhood, revealing instead that the adult brain retains a significant degree of adaptability. This discovery has profound implications for education, rehabilitation, and cognitive enhancement, suggesting that with the right stimuli and training, individuals can reshape their brains to improve memory, learning, and even recovery from injury. The mechanisms underlying neuroplasticity involve complex interactions between neurons, neurotransmitters, and support cells, all working together to remodel the brain's structure and function in response to internal and external stimuli.

4.3.1 Science of Brain Training and Cognitive Enhancement

Brain training, a practice that leverages neuroplasticity to enhance cognitive function, has gained significant attention in both scientific and popular circles.

Unlike passive learning, brain training involves targeted exercises designed to strengthen specific cognitive skills such as memory, attention, problem-solving, and processing speed. These exercises often take the form of digital games, puzzles, or structured mental tasks that challenge the brain to adapt and improve.

Research has shown that consistent engagement in these activities can lead to measurable improvements in cognitive performance, particularly in areas related to working memory and executive function. The effectiveness of brain training is rooted in the principle of "use-dependent plasticity," where the brain strengthens the neural circuits that are frequently activated while pruning those that are underutilized. This process is mediated by changes in synaptic strength, the growth of new dendritic connections, and even neurogenesis in certain brain regions. However, the benefits of brain training are not universal; they depend on the type of training, the individual's baseline cognitive abilities, and the consistency of practice. While some studies have demonstrated significant cognitive gains, others have raised questions about the transferability of these improvements to real-world tasks, highlighting the need for more targeted and contextually relevant training approaches.

4.3.2 Techniques for Harnessing Neuroplasticity

A variety of techniques have been developed to harness neuroplasticity for cognitive enhancement, each leveraging different aspects of the brain's adaptive capacity. One of the most well-established methods is cognitive training, which involves repetitive practice of specific mental tasks to strengthen associated neural pathways. This can include memory exercises, attention-focused games, or problem-solving challenges designed to target particular cognitive domains.

Another powerful approach is dual n-back training, a working memory task that has been shown to improve fluid intelligence by challenging the brain to simultaneously track multiple streams of information. Beyond digital exercises, physical activities such as aerobic exercise and coordination-based movements have been found to enhance neuroplasticity by increasing blood flow to the brain and stimulating the release of growth factors like brain-derived neurotrophic factor (BDNF). Mindfulness and meditation practices also play a significant role, as they have been shown to induce structural changes in brain regions associated with attention, emotional regulation, and self-awareness. Additionally, environmental enrichment—engaging in novel, stimulating activities—can promote neuroplasticity by providing the brain with diverse sensory and cognitive inputs that encourage adaptive reorganization.

4.3.3 Role of Technology in Brain Training

Technology has become an indispensable tool in the field of brain training, offering scalable and accessible platforms for cognitive enhancement. Digital brain training programs, such as those offered by Lumosity, Elevate, and Peak, provide users with a range of exercises designed to target different cognitive functions.

These platforms often incorporate adaptive algorithms that adjust the difficulty of tasks based on user performance, ensuring that the brain is continually challenged at an optimal level. Virtual reality (VR) and augmented reality (AR) are also emerging as powerful tools for brain training, offering immersive environments that can simulate real-world cognitive demands and provide more engaging and contextually relevant training experiences. Wearable neurotechnology, such as EEG headsets and transcranial stimulation devices, allows users to monitor brain activity in real time and even modulate neural function through non-invasive brain stimulation. These technologies not only enhance the effectiveness of brain training but also make it more accessible to a broader audience, enabling individuals to engage in cognitive enhancement from the comfort of their homes. However, the proliferation of brain training technologies also raises questions about their efficacy and the potential for over-reliance on digital solutions at the expense of more holistic cognitive development strategies.

4.3.4 Neuroplasticity in Rehabilitation and Recovery

One of the most impactful applications of neuroplasticity is in the field of rehabilitation, where brain training techniques are used to help individuals recover from neurological injuries and disorders. Stroke survivors, for example, can benefit from constraint-induced movement therapy, a technique that forces the use of affected limbs to stimulate neural reorganization and restore motor function. Similarly, individuals with traumatic brain injuries or neurodegenerative conditions such as Parkinson's disease can engage in targeted cognitive and physical therapies designed to rebuild damaged neural pathways.

The principles of neuroplasticity also underpin the effectiveness of therapies for mental health conditions, including depression and anxiety, where cognitive-behavioral interventions can reshape maladaptive thought patterns and emotional responses. The success of these approaches depends on the brain's ability to adapt and compensate for lost functions, a process that can be enhanced through repetitive, structured, and progressively challenging exercises. While the potential for recovery through neuroplasticity is significant, it is important to note that the extent of rehabilitation varies widely among individuals, depending on factors such as the severity of injury, the timing of intervention, and the consistency of training.

4.3.5 Challenges and Limitations of Brain Training

Despite the promising potential of neuroplasticity and brain training, there are significant challenges and limitations that must be acknowledged. One of the primary concerns is the variability in individual responses to brain training interventions.

While some individuals may experience substantial cognitive improvements, others may see minimal or no benefits, highlighting the complex interplay between genetics, baseline cognitive abilities, and environmental factors. Additionally, the transferability of cognitive gains from training tasks to real-world applications remains a subject of debate, with some studies suggesting that improvements may be limited to the specific tasks practiced rather than generalizing to broader cognitive functions.

There is also the risk of overemphasizing digital brain training at the expense of other critical aspects of cognitive health, such as physical exercise, social interaction, and adequate sleep. Furthermore, the commercialization of brain training programs has led to exaggerated claims and a proliferation of products with questionable scientific backing, making it difficult for consumers to discern effective interventions from marketing hype. Addressing these challenges requires a more nuanced and evidence-based approach to brain training, one that integrates multiple modalities and emphasizes individualized, contextually relevant cognitive enhancement strategies.

4.4 BIOFEEDBACK AND NEUROFEEDBACK IN MENTAL OPTIMIZATION

4.4.1 Biofeedback

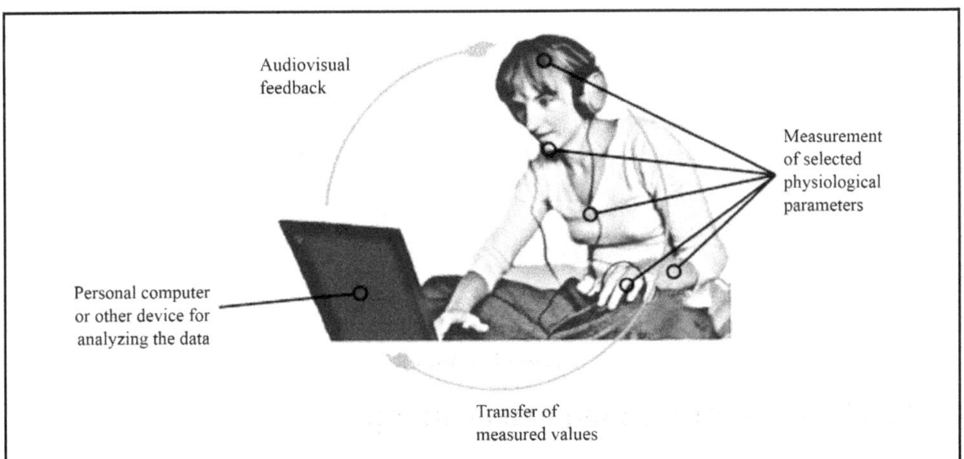

Biofeedback

Biofeedback is a therapeutic and self-regulatory technique that provides individuals with information about their physiological functions. Sensors are placed on the body to measure parameters such as heart rate, skin temperature, muscle tension, or breathing. These signals are then displayed through visual, auditory, or tactile feedback, allowing the user to see how their body responds to stress, relaxation, or focus. With practice, individuals learn to voluntarily control processes that are normally automatic, such as lowering blood pressure or reducing muscle tension. Biofeedback has been widely used in clinical settings to address stress, anxiety, chronic pain, and cardiovascular issues, but it has also become a key tool in the biohacking community for enhancing mental clarity and resilience.

4.4.2 Neurofeedback

Neurofeedback is a specialized form of biofeedback that focuses on brain activity. Using electroencephalography, or EEG, sensors placed on the scalp measure electrical activity in the brain. These signals are processed and displayed in real time, often in the form of graphs, sounds, or interactive games. The goal of neurofeedback is to train individuals to regulate their brainwave patterns, enhancing states of focus, relaxation, or creativity depending on the desired outcome. For example, individuals with attention deficit disorders may learn to increase beta brainwave activity associated with concentration, while those struggling with insomnia may work on promoting alpha and theta waves linked to relaxation. Neurofeedback is increasingly explored as a tool for cognitive enhancement, emotional regulation, and performance optimization.

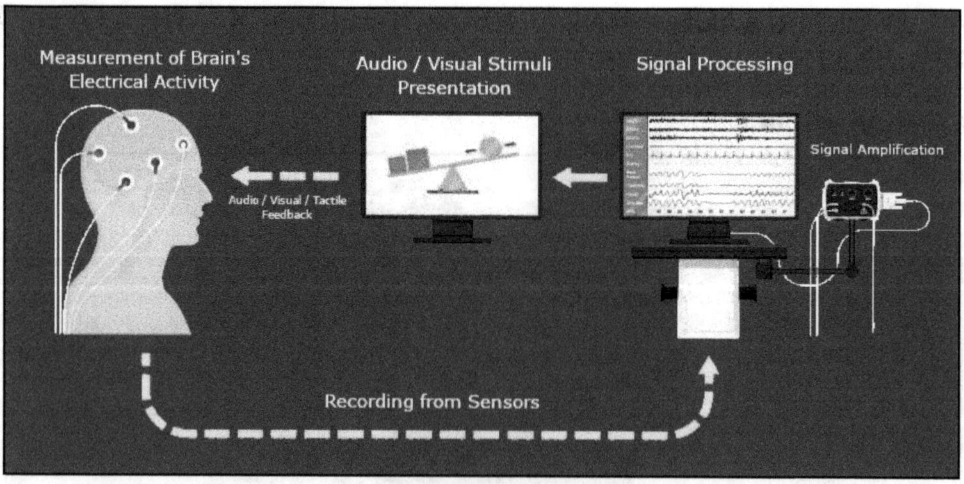

Neurofeedback

4.4.3 Mechanisms of Training and Learning

Both biofeedback and neurofeedback rely on the principle of operant conditioning, where individuals learn to associate specific physiological states with external cues. When users successfully modify their physiological activity toward a desired range, they receive positive feedback, reinforcing the behavior. Over time, this process builds self-awareness and voluntary control over previously unconscious functions. Neuroplasticity, the brain's ability to adapt and reorganize itself, plays a crucial role in neurofeedback, as repeated training sessions can produce lasting changes in brain activity and behavior. This combination of feedback, practice, and adaptation makes biofeedback and neurofeedback unique forms of experiential learning, where the body and mind become active participants in their own optimization.

4.4.4 Applications in Mental Health

One of the most established uses of biofeedback and neurofeedback is in mental health treatment. Biofeedback techniques have been shown to reduce symptoms of anxiety, depression, and post-traumatic stress disorder by teaching individuals to regulate stress responses and promote relaxation. Neurofeedback has been widely studied in attention deficit hyperactivity disorder, where it helps improve focus and reduce impulsivity without reliance on medication. It has also been explored as a therapy for insomnia, addiction, and mood disorders, offering non-invasive alternatives to pharmacological treatments. These applications highlight the therapeutic potential of feedback technologies, providing individuals with tools to manage mental health conditions through self-regulation rather than external intervention.

4.4.5 Cognitive Enhancement and Performance Optimization

Beyond clinical use, biofeedback and neurofeedback have attracted attention as tools for cognitive and performance enhancement. Athletes use heart rate variability biofeedback to improve recovery, resilience, and mental focus under pressure. Musicians and performers employ neurofeedback to enhance flow states, reducing performance anxiety and boosting creativity. Students and professionals experiment with feedback training to improve attention, memory, and productivity. In these contexts, the technologies are less about treating dysfunction and more about reaching peak mental performance. The appeal lies in the ability to fine-tune the mind, training it as one would train a muscle, to respond optimally in demanding situations.

4.4.6 Integration with Technology and Wearables

Recent advances in technology have made biofeedback and neurofeedback more accessible to the general public. Portable devices, smartphone applications, and wearable sensors allow users to track physiological signals such as heart rate variability, skin conductance, or brainwaves outside laboratory or clinical settings. Consumer-grade EEG headbands, for example, provide real-time neurofeedback for meditation and focus training. Similarly, wearable heart rate monitors enable individuals to practice breathing techniques for stress reduction. These developments align with the biohacking ethos of self-quantification and experimentation, making feedback technologies part of everyday routines rather than specialized therapies. The integration of biofeedback with wearables also enables long-term tracking, offering deeper insights into personal progress and patterns.

4.4.7 Challenges and Limitations

Despite their promise, biofeedback and neurofeedback face challenges and limitations that temper their widespread adoption. The effectiveness of these methods can vary significantly between individuals, with some showing marked improvement and others experiencing little benefit. Access to high-quality equipment and professional guidance is often necessary for optimal results, though consumer devices attempt to bridge this gap. In neurofeedback especially, the interpretation of brainwave patterns is complex, and oversimplification in consumer products risks producing misleading outcomes. Scientific research supports many applications of biofeedback, but neurofeedback remains a field with ongoing debate and mixed evidence, requiring more rigorous studies. These limitations highlight the importance of cautious adoption, balancing enthusiasm with realistic expectations.

4.4.8 Ethical and Societal Implications

As with other forms of cognitive enhancement, biofeedback and neurofeedback raise ethical and societal questions. In clinical contexts, they offer valuable non-invasive alternatives, but in performance or cognitive optimization, they blur lines between therapy and enhancement. Concerns about fairness, accessibility, and potential misuse emerge, particularly if neurofeedback becomes a tool for competitive advantage in academic or professional settings. Privacy also poses an issue, as brainwave data and physiological markers are deeply personal. The increasing commercialization of neurofeedback devices raises questions about regulation, quality control, and responsible marketing. Society must grapple with these ethical challenges to ensure that feedback technologies are used responsibly and inclusively.

4.4.9 Role in the Biohacking Movement

Within the broader biohacking movement, biofeedback and neurofeedback represent some of the most sophisticated tools for mental optimization. Unlike passive supplements or devices, feedback systems actively engage individuals in training their bodies and minds. Biohackers value them for their combination of quantification, control, and self-experimentation, using them alongside meditation, nootropics, or sleep optimization strategies. The culture of experimentation and community sharing in biohacking accelerates the exploration of feedback technologies, often pushing boundaries beyond traditional medical or academic frameworks. This grassroots exploration reflects both the promise and the risks of democratized mental optimization, where individuals experiment with their cognitive potential using accessible yet powerful tools.

4.5 SLEEP AS A COGNITIVE BIOHACK

In the relentless pursuit of cognitive enhancement, biohackers often gravitate toward sophisticated technologies and exotic supplements, overlooking the most powerful and fundamental performance tool available: sleep. Far from a passive state of inactivity, sleep is an intense period of biological alchemy, a non-negotiable process of restoration, consolidation, and cleansing that forms the very foundation of mental acuity. Framing sleep as a cognitive biohack shifts the perspective from one of mere rest to one of active, strategic investment in brain function. It is the ultimate leverage point, the baseline upon which all other cognitive enhancements are built; without quality sleep, no amount of nootropics or neurofeedback can achieve its full potential. Optimizing sleep is not about spending more time in bed, but about engineering the conditions for higher-quality, more efficient sleep to maximize its restorative payoff for the mind.

4.5.1 Neurobiology of the Nightshift: What Happens to the Brain Asleep

To harness sleep as a biohack, one must first understand the intricate cognitive work being performed during the night. Sleep is not a monolithic state but a cyclical journey through distinct stages, each with a specialized function for the brain. The first critical phase is deep non-REM sleep, or slow-wave sleep. This stage is characterized by synchronized, slow-frequency brainwaves and is paramount for memory consolidation. It is during deep sleep that the hippocampus, the brain's temporary storage site for new memories, replays the day's events, transferring them to the neocortex for long-term storage, effectively transforming fragile short-term memories into stable, lasting knowledge.

Following deep sleep, the brain enters REM (Rapid Eye Movement) sleep. Here, the brain is almost as active as when awake, but the body is paralyzed. REM sleep is the stage most associated with vivid dreaming and is crucial for emotional regulation and creative problem-solving. It is during REM that the brain disconnects and reassembles memory fragments, forging novel connections between disparate ideas, which is the bedrock of creativity and insight. Concurrently, a recently discovered system called the glymphatic system becomes highly active during deep sleep. This waste-clearance system flushes the brain with cerebrospinal fluid, washing away metabolic debris and toxic proteins, such as beta-amyloid, which are associated with neurodegenerative diseases. This nightly cleanse is essential for maintaining cognitive clarity and long-term brain health.

4.5.2 Engineering the Environment: The Foundation of Sleep Hygiene

The first and most effective strategy for hacking sleep quality is the deliberate optimization of the sleep environment, a practice known as sleep hygiene. This involves creating a cave-like setting that is cool, dark, and quiet. Temperature is a critical regulator of the sleep-wake cycle; a drop in core body temperature is a key signal to initiate sleep. Maintaining a bedroom temperature between 60-67 degrees Fahrenheit (15-19 degrees Celsius) supports this natural thermoregulation. Absolute darkness is equally vital, as even small amounts of light, particularly blue light from electronic devices, can suppress the production of melatonin, the hormone that cues sleepiness.

This necessitates the use of blackout curtains, the removal of all LED indicators, and the strict avoidance of screens for at least an hour before bed. For those who cannot control external noise, white noise machines or apps can mask disruptive sounds with a consistent, soothing auditory blanket. The bed itself is a primary tool; investing in a high-quality mattress and pillows that support proper spinal alignment can drastically reduce sleep-disrupting tosses and turns. This deliberate engineering of the external environment removes barriers to sleep initiation and maintenance, allowing the natural sleep architecture to unfold without interruption.

4.5.3 Chronobiology and Rhythm Regulation: Hacking the Circadian Clock

Beyond the immediate environment, sleep is governed by the body's master clock, the suprachiasmatic nucleus, which aligns our physiology with the 24-hour solar day. Hacking this circadian rhythm is perhaps the most powerful lever for achieving consistent, restorative sleep. The primary cue for this system is light exposure. Viewing bright, natural light within the first hour of waking, ideally through outdoor exposure, strongly signals the clock to suppress melatonin and promote alertness, setting a clear countdown for its release approximately 14-16 hours later.

Conversely, as evening approaches, minimizing exposure to blue and green spectrum light is crucial. This can be achieved through software like f.lux or smartphone night shift modes, and by wearing blue-light-blocking glasses. The timing of food intake also influences the circadian rhythm. Establishing a consistent feeding window, such as finishing all meals 2-3 hours before bedtime, allows the body to focus on repair during sleep rather than digestion. Furthermore, maintaining a rigid sleep schedule—waking up and going to bed at the same time every day, including weekends—is the single most effective habit for reinforcing a robust circadian rhythm, training the body to anticipate and prepare for sleep with precision.

4.5.4 Nutritional and Supplemental Synergy: Fueling the Sleep Cycle

Nutrition provides the raw materials for sleep biochemistry, and strategic choices can significantly enhance sleep quality and architecture. The amino acid tryptophan, found in foods like turkey, dairy, and nuts, is a precursor to both serotonin and melatonin. Consuming a small tryptophan-containing snack a few hours before bed can support this natural production process. Minerals also play a key role; magnesium glycinate is a highly bioavailable form of magnesium that acts as a natural NMDA receptor antagonist and GABA agonist, promoting muscle relaxation and nervous system calm.

Similarly, the mineral zinc is involved in the conversion of tryptophan to melatonin. For those seeking supplemental aid, melatonin itself can be effective for resetting the sleep cycle in cases of jet lag or shift work, but its chronic use is debated. More sophisticated nootropics for sleep include apigenin, a flavonoid found in chamomile that acts as a positive allosteric modulator at GABA-A receptors, and low doses of theanine, which can promote alpha brain waves associated with relaxed alertness and improve sleep quality. It is critical to avoid substances that disrupt sleep architecture, most notably alcohol, which suppresses REM sleep, and caffeine, whose long half-life can linger in the system and impair sleep quality even if it does not prevent sleep onset.

4.5.5 Cognitive Dividend of Optimized Sleep

The payoff for investing in sleep optimization is immense and directly impacts every facet of cognitive performance. A well-rested brain demonstrates superior executive function: enhanced focus, sharper decision-making, and greater emotional resilience. The memory consolidation that occurs during deep sleep directly translates to improved learning efficiency, making study and skill acquisition more effective. The creative insights forged during REM sleep provide solutions to complex problems and foster innovation.

Furthermore, the nightly glymphatic cleanse ensures that cognitive function remains sharp and clear, free from the mental fog that arises from metabolic buildup. By prioritizing and systematically optimizing sleep, the biohacker is not merely preventing fatigue; they are actively upgrading their brain's operational software every single night. It is the ultimate cognitive biohack because it is foundational, synergistic, and free. In the economy of human performance, sleep is the highest-yield investment, compounding its benefits each night to build a stronger, faster, and more resilient mind.

CHAPTER 5

Physical Biohacking and Longevity

5.1 HORMESIS AND ADAPTATION-BASED TRAINING

In the quest for extended healthspan and vitality, biohackers and longevity enthusiasts increasingly focus on methods that push the body beyond its comfort zone in controlled ways. One of the central principles underlying these practices is hormesis, the biological phenomenon where exposure to mild stressors triggers adaptive responses that make the body more resilient.

Coupled with adaptation-based training, which emphasizes progressive challenges to physical systems, hormesis represents a powerful framework for physical biohacking. By intentionally applying stressors such as fasting, exercise, temperature extremes, or hypoxia, individuals aim to enhance their resilience, slow aging, and promote longevity. Understanding the science and philosophy of hormesis reveals how strategic stress can be a pathway to stronger, healthier, and longer lives.

5.1.1 Hormesis

Hormesis is a biological principle rooted in toxicology and biology, describing a biphasic response to stress or exposure. At high levels, a stressor can be damaging or lethal, but at low or moderate levels, it can stimulate beneficial adaptations. For instance, radiation, toxins, or exercise can harm the body in excess, yet in smaller doses they activate cellular repair mechanisms and promote resilience. This paradox illustrates the concept of "what does not kill you makes you stronger," but in a precise biological sense. In the context of biohacking, hormesis is harnessed intentionally, where the stressor is carefully controlled to avoid harm while maximizing adaptive benefits.

5.1.2 Adaptation and Resilience

Adaptation is the body's natural response to repeated stress. When faced with a challenge, whether physical, metabolic, or environmental, the body mobilizes resources to cope. If the challenge is not overwhelming, the system adapts, becoming stronger and better equipped to handle future stress. This principle is evident in physical training, where muscles grow stronger after being subjected to resistance or in immunity, where exposure to pathogens builds defense. Adaptation-based training builds upon this principle, deliberately designing stress exposures to optimize resilience. The focus is not on avoiding stress but on embracing it in measured ways, training the body to withstand greater demands over time.

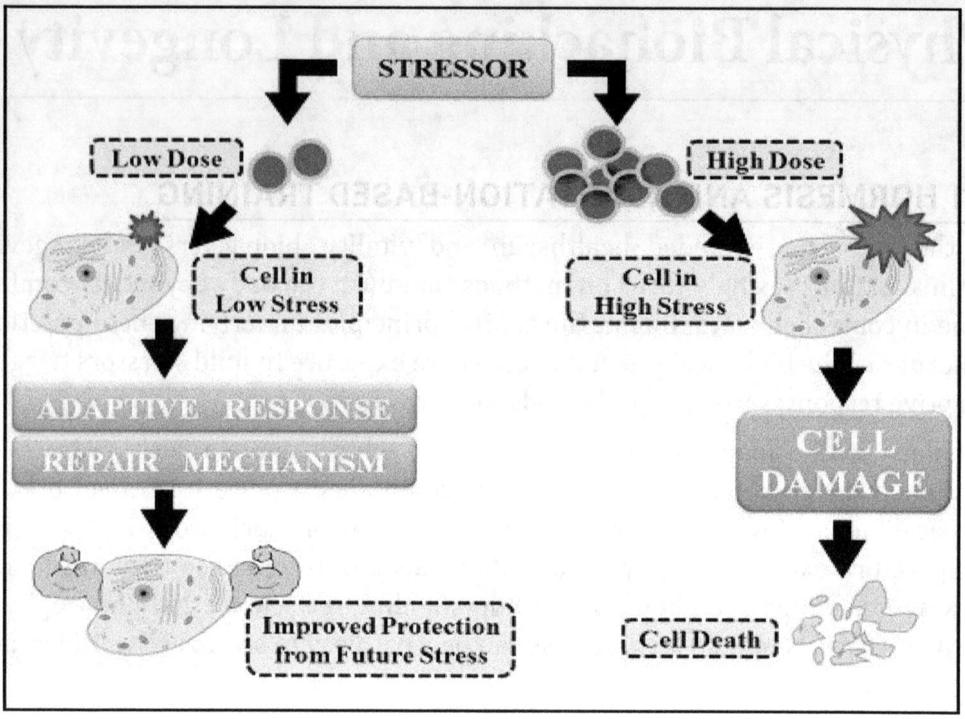

Hormesis

5.1.3 Exercise as a Model of Hormesis

Exercise is the most widely recognized form of hormesis in action. When individuals engage in resistance training, cardiovascular workouts, or high-intensity intervals, they create microdamage, oxygen stress, and energy depletion. These stressors trigger adaptations such as muscle repair, mitochondrial biogenesis, and improved cardiovascular capacity. Overtraining can be harmful, but moderate, repeated exercise makes the body stronger and healthier. This hormetic effect of exercise explains why physical activity is so strongly associated with longevity. From endurance sports that enhance cardiovascular efficiency to strength training that preserves muscle mass in aging, exercise illustrates how controlled stress drives long-term resilience.

5.1.4 Fasting and Metabolic Stress

Another important form of hormesis is fasting, which introduces metabolic stress by depriving the body of nutrients for a period of time. During fasting, cells shift into repair and maintenance mode, activating pathways such as autophagy, which recycles damaged cellular components. Short-term nutrient stress stimulates insulin sensitivity, reduces inflammation, and enhances metabolic flexibility. While chronic starvation is harmful, intermittent fasting and caloric restriction mimic hormetic effects, contributing to healthspan and longevity. Biohackers experiment with various fasting protocols, such as time-restricted feeding, alternate-day fasting, or prolonged fasts, to activate adaptive responses that promote cellular health and delay aging processes.

5.1.5 Temperature Stress: Heat and Cold Exposure

Exposure to extreme temperatures provides another form of hormesis that has gained popularity in the biohacking community. Heat exposure, as in saunas or hot baths, triggers heat-shock proteins that protect cellular integrity and promote stress resistance. Cold exposure, through ice baths, cold showers, or cryotherapy, activates brown adipose tissue, enhances circulation, and improves resilience to stress. Both forms of temperature stress induce adaptations that improve recovery, reduce inflammation, and support longevity. The deliberate use of thermal stress demonstrates how environmental factors can be transformed into tools for building resilience through controlled hormetic exposure.

5.1.6 Hypoxia and Breathwork

Hypoxia, or low oxygen availability, is another stressor that can be used in adaptation-based training. Practices such as altitude training, breath-hold exercises, or intermittent hypoxic training push the body to function with reduced oxygen. In response, the body increases red blood cell production, enhances oxygen efficiency, and stimulates pathways that protect against oxidative stress. Breathwork practices, including those inspired by ancient traditions or modern techniques like the Wim Hof Method, combine controlled breathing with hypoxia and cold exposure to create powerful adaptive responses. These practices illustrate how nontraditional stressors can contribute to resilience and longevity through hormesis.

5.1.7 Molecular Mechanisms of Hormesis

The benefits of hormesis are underpinned by cellular and molecular mechanisms. Stressors such as exercise, fasting, or thermal exposure activate pathways like AMP-activated protein kinase, sirtuins, and the Nrf2 antioxidant system. These pathways regulate energy metabolism, DNA repair, and cellular defense against oxidative damage. Autophagy, mitochondrial biogenesis, and the upregulation of heat-shock proteins are central processes that enhance resilience and slow cellular aging. Hormesis essentially activates the body's repair and maintenance systems, which tend to decline with age. By regularly triggering these pathways through mild stress, individuals can counteract aspects of biological aging.

5.1.8 Hormesis and Longevity Research

Research on hormesis has increasingly linked stress-induced adaptation with lifespan extension in various species. Studies on caloric restriction, exercise, and heat stress in animals consistently show increases in lifespan and healthspan. In humans, long-term observational studies highlight associations between hormetic practices and reduced risks of chronic diseases such as cardiovascular illness, diabetes, and neurodegeneration. While direct evidence for lifespan extension in humans is harder to establish, the accumulation of data suggests that hormesis-based interventions contribute to resilience, vitality, and slower biological aging. Longevity science increasingly views mild stressors not as threats but as tools for health optimization.

5.1.9 Risks and Misapplication of Hormesis

While hormesis holds great promise, it must be applied with care. The boundary between beneficial stress and harmful overload is often narrow, and overexposure to stressors can lead to injury, exhaustion, or chronic damage. Overtraining in exercise, excessive fasting, or prolonged cold exposure can undermine health rather than enhance it. Individual variability also plays a role, as genetics, age, and underlying health conditions affect how one responds to stress. Misunderstanding hormesis as a justification for extreme practices can be dangerous, underscoring the need for balance and moderation. Adaptation-based training works best when stressors are applied progressively, with sufficient recovery to allow beneficial adaptation.

5.1.10 Role of Biohacking in Hormetic Practices

Within the biohacking community, hormesis has become a guiding principle for physical and lifestyle experimentation. Biohackers adopt stress-based interventions in combination with modern tracking tools, such as heart rate variability monitors, glucose sensors, or recovery wearables, to measure and refine their responses. Cold plunges, fasting windows, and high-intensity training regimens are shared practices in biohacking circles, often combined with data-driven personalization. The integration of hormesis into biohacking reflects a shift from seeking comfort and ease toward embracing calculated discomfort as a route to growth and longevity.

5.1.11 Philosophical Dimensions of Stress and Growth

The philosophy behind hormesis and adaptation-based training extends beyond biology into how individuals approach life. It challenges the notion that health comes from avoiding stress and instead emphasizes resilience through controlled struggle. This mindset aligns with broader human traditions, from Spartan training to ascetic practices that valued endurance and self-mastery. In modern contexts, hormesis provides a scientific justification for embracing discomfort, whether through fasting, exertion, or environmental extremes. The idea resonates with a philosophy of growth through challenge, shaping not only physical longevity but also mental toughness and adaptability.

5.2 EXERCISE SCIENCE AND RECOVERY BIOHACKS

5.2.1 Science of Stimulus: Principles of Exercise for Adaptation

Effective physical biohacking begins with understanding the different levers of exercise stress and how to pull them to achieve specific outcomes. The foundational principle is that of hormesis, the concept that a moderate, acute stressor induces a beneficial adaptive response, while the same stressor at a chronic or extreme level becomes damaging. Exercise is the quintessential hormetic stressor. Resistance training, through mechanical tension and metabolic stress, induces micro-damage in muscle fibers, signaling the body to not only repair but overcompensate by building stronger, larger muscles—a process known as hypertrophy.

This process is crucial for longevity, as it combat sarcopenia, the age-related loss of muscle mass that is a primary driver of frailty. High-Intensity Interval Training (HIIT) provides a potent cardiorespiratory and metabolic stimulus. By alternating short bursts of all-out effort with periods of active recovery, HIIT improves mitochondrial density and efficiency, enhances insulin sensitivity, and boosts VO2 max, a key marker of aerobic fitness and longevity, often in less time than steady-state cardio. Low-intensity steady-state (LISS) cardio, such as brisk walking or cycling, supports recovery by promoting blood flow without imposing significant systemic stress, thereby aiding in the clearance of metabolic waste and reducing inflammation. A biohacker's exercise regimen is not random but is a periodized program that strategically alternates between these different modalities to continuously challenge the body while avoiding the plateauing and overtraining that hinder long-term progress.

5.2.2 Crucible of Growth: Why Recovery is the True Biohack

If exercise is the question, recovery is the answer. The body's adaptive processes are entirely dependent on the resources and environment provided during the rest period. Inadequate recovery leads to a state of overreaching, which can quickly devolve into overtraining syndrome, characterized by hormonal imbalances, suppressed immune function, persistent fatigue, and decreased performance. Therefore, the modern approach to physical enhancement dedicates at least equal attention to the science of recuperation. Recovery is multifaceted, encompassing physiological, neurological, and psychological restoration. Physiologically, it involves the repair of damaged tissues, the replenishment of glycogen stores in muscles and the liver, and the rebalancing of the hormonal milieu, particularly cortisol and testosterone. Neurologically, it requires the recalibration of the central nervous system, which can become fatigued from coordinating high-intensity muscular contractions. Psychologically, it necessitates a reduction in mental fatigue and stress, which can themselves impair physical performance and slow healing. Viewing recovery as an active process to be optimized, rather than a passive state of inactivity, is the core tenet of physical biohacking for longevity.

5.2.3 Technological and Nutritional Tools for Enhanced Recuperation

A suite of technologies and nutritional strategies has emerged to accelerate and enhance the body's innate recovery processes. Among the most popular technological tools are percussion therapy devices, such as the Theragun, which use rapid, targeted pulses to manipulate deep tissue, increasing blood flow, reducing muscle stiffness, and potentially breaking up adhesions. Whole-Body Cryotherapy, which involves brief exposure to extremely cold air (-200 to -300°F) in a specialized chamber, aims to reduce inflammation and muscle soreness by constricting blood vessels and flushing the area with nutrient-rich blood upon rewarming.

Similarly, infrared saunas use light to create heat that penetrates deep into tissues, promoting detoxification through sweat, increasing circulation, and stimulating the release of endorphins. Contrast Water Therapy, alternating between hot and cold water immersion, exploits the body's vascular response to pump fluids and reduce inflammation. Nutritionally, recovery is supported by targeted nutrient timing. Consuming a combination of fast-digesting protein and carbohydrates within the "anabolic window" following exercise provides the essential amino acids for muscle protein synthesis and the glucose to rapidly replenish glycogen stores. Supplements like branched-chain amino acids (BCAAs) can further support muscle repair, while electrolytes are crucial for rehydration and optimal neuromuscular function. Omega-3 fatty acids, found in fish oil, are potent anti-inflammatories that can modulate the inflammatory response to training, and tart cherry juice has been shown to reduce muscle soreness and speed strength recovery due to its high antioxidant and anti-inflammatory content.

5.2.4 Longevity Dividend: Integrating Exercise and Recovery for Lifelong Health

The ultimate goal of integrating precision exercise with advanced recovery biohacks is not merely short-term performance enhancement but the long-term extension of healthspan—the period of life spent in good health. The synergistic effects of this approach are profound. Maintaining muscle mass through resistance training is one of the strongest predictors of longevity, as it regulates metabolism, supports immune function, and preserves functional independence. Improving cardiovascular health through HIIT and LISS directly reduces the risk of all-cause mortality, heart disease, and stroke. The deliberate management of inflammation through recovery protocols is perhaps the most critical factor, as chronic, systemic inflammation is a primary driver of nearly every age-related disease, from arthritis and diabetes to neurodegeneration and cancer. By consistently applying a hormetic exercise stress and then actively facilitating a robust recovery, the biohacker creates a positive feedback loop of adaptation. The body becomes increasingly resilient, efficient, and capable of resisting the degradative processes of aging. This approach reframes fitness from an aesthetic pursuit to a fundamental longevity strategy. It acknowledges that the body is a complex system that thrives on challenge and flourishes with rest, and that the conscious application of both is the most powerful physical biohack available for building a longer, healthier, and more vibrant life.

5.3 ANTI-AGING INTERVENTIONS

5.3.1 Senolytics, NAD+, and Rapamycin

Senolytics: Clearing the Way for Cellular Rejuvenation

One of the most promising areas of anti-aging research involves senolytics, a class of compounds designed to selectively eliminate senescent cells—cells that have stopped dividing and accumulate with age, contributing to inflammation, tissue dysfunction, and age-related diseases. Senescent cells secrete a range of pro-inflammatory molecules known as the senescence-associated secretory phenotype (SASP), which can damage surrounding tissues and accelerate aging. Senolytics work by targeting pathways that allow these cells to evade programmed cell death, effectively clearing them from the body and reducing chronic inflammation. Early research in animal models has shown that senolytic treatments can reverse aspects of aging, improve physical function, and extend lifespan. Compounds such as dasatinib and quercetin, fisetin, and a combination of navitoclax and other agents have demonstrated senolytic effects in preclinical studies, with some now being tested in human clinical trials. The potential of senolytics lies not just in extending lifespan but in improving healthspan by delaying or mitigating age-related conditions such as osteoarthritis, cardiovascular disease, and neurodegenerative disorders. However, the long-term effects of senolytic therapies in humans remain to be fully understood, and their use in biohacking circles is still largely experimental, requiring careful consideration of risks and benefits.

NAD+ Boosting: Replenishing the Fuel of Cellular Youth

Nicotinamide adenine dinucleotide (NAD+) is a critical coenzyme found in every cell of the body, playing a central role in energy metabolism, DNA repair, and the regulation of cellular aging. As we age, NAD+ levels decline, contributing to mitochondrial dysfunction, impaired cellular repair, and increased susceptibility to age-related diseases. This has led to significant interest in NAD+ precursors such as nicotinamide riboside (NR) and nicotinamide mononucleotide (NMN), which can be supplemented to restore NAD+ levels and potentially reverse aspects of cellular aging. Research suggests that boosting NAD+ can enhance mitochondrial function, improve metabolic health, and even extend lifespan in animal models. In humans, NAD+ augmentation has been associated with improvements in muscle function, cognitive performance, and cardiovascular health.

The growing availability of NR and NMN supplements has made NAD+ boosting a popular intervention in the biohacking community, though the optimal dosing, long-term safety, and efficacy in humans are still areas of active investigation. The potential of NAD+ to act as a metabolic rejuvenator makes it a cornerstone of modern anti-aging strategies, offering a pathway to restore cellular vitality and delay the onset of age-related decline.

Rapamycin: The Longevity Drug with Immunosuppressive Roots

Rapamycin, originally developed as an immunosuppressant and later recognized for its anti-cancer properties, has emerged as one of the most intriguing compounds in the field of longevity research. Its ability to inhibit the mechanistic target of rapamycin (mTOR), a key regulator of cell growth, metabolism, and aging, has positioned it as a leading candidate for extending healthspan and lifespan. By dampening mTOR activity, rapamycin mimics the effects of caloric restriction, a well-established intervention for promoting longevity, and enhances cellular autophagy—the process by which cells remove damaged components and recycle nutrients.

Studies in various organisms, from yeast to mammals, have demonstrated that rapamycin can extend lifespan and delay the onset of age-related diseases. However, its use in humans is complicated by its immunosuppressive effects, which raise concerns about long-term safety and potential side effects. This has led to interest in rapamycin analogs, or "rapalogs," such as everolimus, which may offer similar longevity benefits with fewer adverse effects. The biohacking community has shown significant interest in rapamycin, though its use remains largely experimental, with ongoing research needed to determine optimal dosing and long-term safety profiles. The potential of rapamycin to modulate fundamental aging pathways makes it a compelling, albeit complex, tool in the pursuit of extended healthspan.

Integrating Anti-Aging Interventions: A Holistic Approach

While senolytics, NAD+ boosters, and rapamycin each offer unique mechanisms for targeting aging, their integration into a comprehensive longevity strategy requires careful consideration of individual biology, health status, and goals. The biohacking community often explores these interventions in combination, seeking synergistic effects that might amplify their benefits. For example, combining senolytics to clear dysfunctional cells with NAD+ supplementation to restore metabolic function could theoretically enhance cellular rejuvenation more effectively than either intervention alone.

Similarly, the use of rapamycin to promote autophagy might be complemented by NAD+ augmentation to support mitochondrial health, creating a multi-faceted approach to aging. However, such combinations must be approached with caution, as the interactions between these compounds are not yet fully understood, and their long-term effects in humans remain under investigation. The key to successful integration lies in a measured, data-driven approach, where interventions are carefully monitored, adjusted, and evaluated for efficacy and safety. This holistic perspective emphasizes not just the use of specific compounds but the broader context of lifestyle, nutrition, and environmental factors that collectively influence aging and longevity.

Challenges and Ethical Considerations in Anti-Aging Biohacking

The pursuit of longevity through interventions like senolytics, NAD+, and rapamycin is not without challenges and ethical considerations. One of the primary concerns is the lack of long-term human data, as much of the research supporting these interventions comes from animal models or short-term clinical trials. This raises questions about the safety, efficacy, and potential unintended consequences of their use in healthy individuals seeking to extend healthspan. Additionally, the accessibility of these interventions varies widely, with some compounds being readily available as supplements while others remain prescription-only or experimental, creating disparities in who can benefit from these advancements. Ethical dilemmas also arise in the context of biohacking, where individuals may engage in self-experimentation without medical supervision, potentially exposing themselves to risks. The commercialization of anti-aging interventions further complicates the landscape, as marketing claims can sometimes outpace scientific evidence, leading to misinformation and unrealistic expectations. Navigating these challenges requires a balanced approach, where curiosity and innovation are tempered by scientific rigor, transparency, and a commitment to safety. The responsible use of these interventions necessitates ongoing research, open dialogue, and a recognition of the complexities inherent in manipulating fundamental biological processes.

5.3.2 Telomere Testing and Biological Age Measurement

Telomeres: The Cellular Clock of Aging

At the heart of biological age measurement lies the study of telomeres, the protective caps at the ends of chromosomes that shorten with each cell division. Telomeres act as a biological clock, with their length serving as a key indicator of cellular aging and overall health. As cells replicate, telomeres gradually erode, eventually reaching a critical length that triggers cellular senescence or apoptosis.

This process is influenced by various factors, including oxidative stress, inflammation, and lifestyle choices, making telomere length a powerful biomarker for biological aging. Research has shown that individuals with longer telomeres tend to exhibit better health outcomes and increased longevity, while accelerated telomere shortening is associated with age-related diseases and premature aging.

The discovery of telomerase, an enzyme that can lengthen telomeres, has sparked interest in potential anti-aging interventions, though its therapeutic application remains complex and largely experimental. For biohackers and longevity enthusiasts, telomere testing offers a tangible metric to gauge the impact of lifestyle modifications, supplements, and other interventions on cellular aging.

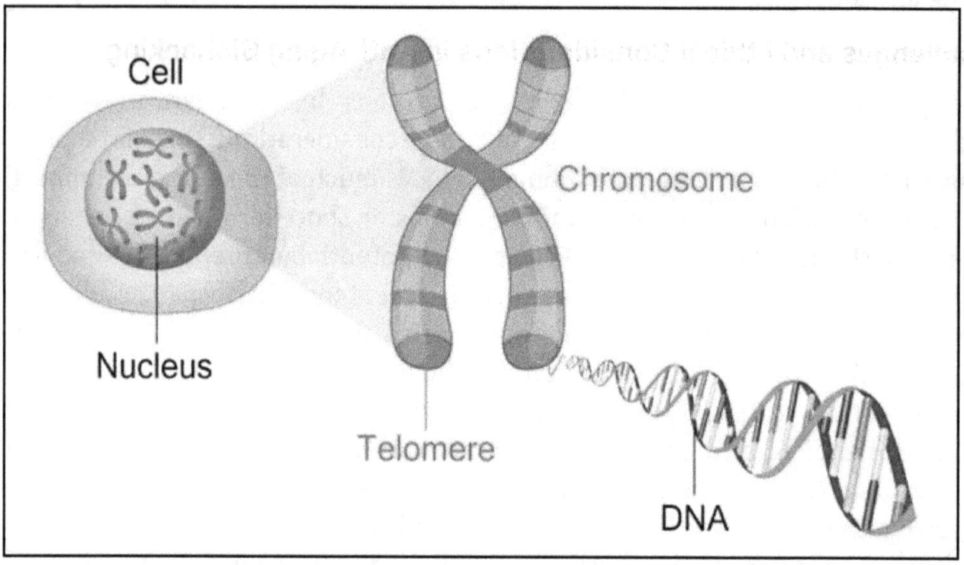

Telomeres

Methods of Measuring Biological Age

The measurement of biological age has evolved significantly with advancements in biotechnology, offering multiple approaches to assess physiological aging. One of the most well-established methods is telomere length analysis, typically performed through polymerase chain reaction (PCR) or fluorescence in situ hybridization (FISH) techniques on DNA samples. These tests provide a snapshot of cellular aging by quantifying the average length of telomeres in white blood cells. Another prominent approach involves epigenetic clocks, which analyze DNA methylation patterns—chemical modifications to DNA that change with age—to predict biological age with remarkable accuracy. The Horvath clock and other epigenetic models have become gold standards in aging research, offering insights into how lifestyle factors influence aging at the molecular level. Additionally, composite biomarker panels that assess factors such as inflammation markers, metabolic function, and cardiovascular health provide a more holistic view of biological age. These measurements, often available through direct-to-consumer testing services, empower individuals to track their aging trajectory and make informed decisions about health optimization strategies.

Role of Telomere Testing in Longevity Biohacking

Telomere testing has become a cornerstone of longevity biohacking, providing a concrete metric to evaluate the effectiveness of anti-aging interventions. For biohackers, knowing their telomere length offers a baseline from which to assess the impact of dietary changes, exercise regimens, stress management techniques, and supplement protocols. Interventions such as caloric restriction, antioxidant-rich diets, and mindfulness practices have been shown to slow telomere shortening, while chronic stress, poor sleep, and unhealthy lifestyle habits can accelerate it. Some biohackers explore more experimental approaches, including the use of telomerase-activating compounds like TA-65, though the long-term safety and efficacy of such interventions remain subjects of ongoing research. The ability to measure telomere length over time allows for a personalized approach to aging, where individuals can adjust their strategies based on empirical data rather than anecdotal advice. However, it is important to note that telomere length is just one piece of the aging puzzle, and its interpretation should be considered alongside other biomarkers and health metrics for a comprehensive understanding of biological age.

Epigenetic Clocks and the Future of Aging Measurement

Epigenetic clocks represent one of the most exciting advancements in the field of biological age measurement, offering a sophisticated and highly accurate method to assess aging at the molecular level. These clocks work by analyzing specific sites on the genome where DNA methylation—a process that regulates gene expression—changes predictably with age. The Horvath clock, for example, can estimate biological age across various tissues and cell types, providing insights into how different organs and systems are aging relative to one another. This level of precision allows for a more nuanced understanding of aging, revealing how lifestyle factors, environmental exposures, and even psychological stress can influence the epigenetic landscape. The development of second-generation epigenetic clocks, which incorporate additional biomarkers and machine learning algorithms, promises even greater accuracy and predictive power. For biohackers, these tools offer an unprecedented opportunity to track aging in real time, enabling highly personalized interventions that target specific biological pathways. As research continues to refine these clocks, their integration into consumer health platforms will likely make biological age measurement a standard component of preventive healthcare and longevity optimization.

Practical Applications of Biological Age Testing

The practical applications of biological age testing extend far beyond mere curiosity, offering tangible benefits for health optimization and disease prevention. Individuals who undergo telomere or epigenetic testing can use the results to guide lifestyle modifications, such as adopting diets rich in antioxidants and anti-inflammatory compounds, optimizing sleep patterns, and incorporating stress-reducing practices like meditation and exercise. For those engaged in more advanced biohacking, biological age metrics can inform the use of supplements like NAD+ boosters, senolytics, or rapamycin analogs, providing a way to monitor their effects on cellular aging. Additionally, biological age testing can serve as a motivational tool, offering concrete evidence of how positive lifestyle changes translate into measurable improvements in health and longevity. Some longevity clinics and wellness programs now incorporate biological age testing as part of comprehensive health assessments, using the data to tailor personalized interventions. As these tests become more accessible and affordable, their role in preventive medicine and personalized health optimization will continue to grow, empowering individuals to take proactive control of their aging process.

Challenges and Ethical Considerations in Biological Age Measurement

Despite the promise of biological age measurement, there are significant challenges and ethical considerations that must be addressed. One of the primary concerns is the potential for misinterpretation or overreliance on biological age metrics, where individuals might make drastic health decisions based on incomplete or misunderstood data. The accuracy and reproducibility of telomere and epigenetic tests can vary, and their results should be interpreted within the broader context of an individual's health profile. Additionally, the commercialization of biological age testing raises questions about data privacy, as genetic and epigenetic information is highly sensitive and could be vulnerable to misuse. Ethical dilemmas also arise in the context of biohacking, where individuals might pursue unproven or risky interventions based on their biological age results, potentially exposing themselves to harm. The accessibility of these tests also highlights disparities in who can afford and benefit from advanced longevity technologies, raising concerns about equity in health optimization. Navigating these challenges requires a balanced approach, where biological age measurement is used as one tool among many in a comprehensive, evidence-based strategy for health and longevity.

5.4 MITOCHONDRIAL HEALTH AND ENERGY OPTIMIZATION

Mitochondria, often referred to as the powerhouses of the cell, are central to human health, performance, and longevity. These tiny organelles are responsible for generating the energy that fuels every biological process, from muscle contraction to brain activity. Beyond their role in energy production, mitochondria influence cellular signaling, immunity, and the regulation of aging. In the context of biohacking and human optimization, mitochondrial health has emerged as a crucial focus, as optimizing mitochondrial function can enhance vitality, slow the aging process, and improve resilience against disease.

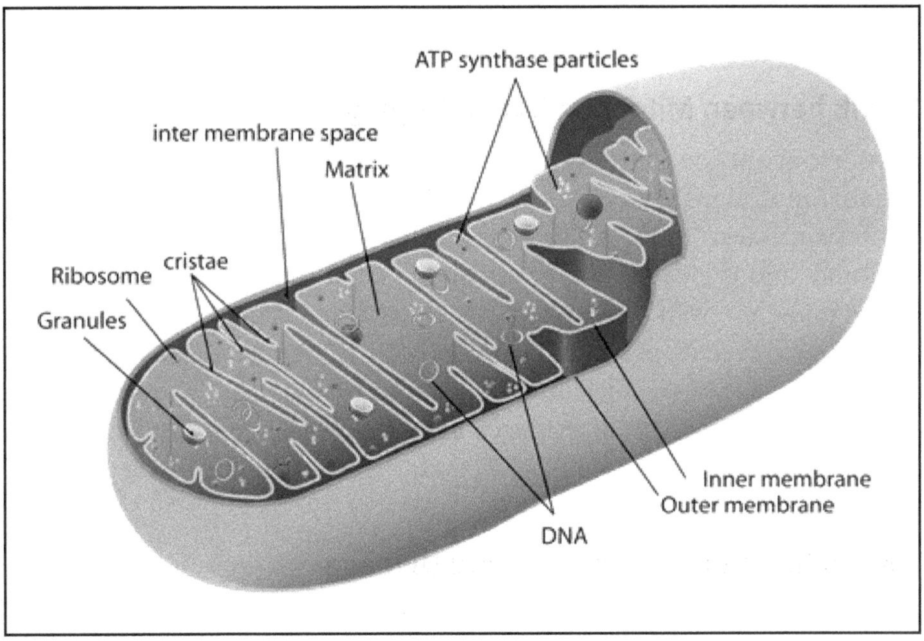

Mitochondria

5.4.1 Role of Mitochondria in Energy Production

Mitochondria are unique organelles that generate adenosine triphosphate, or ATP, the universal currency of energy in biological systems. They achieve this through oxidative phosphorylation, a process in which electrons derived from nutrients such as glucose and fatty acids pass through the electron transport chain. This transfer of electrons creates a proton gradient that powers the synthesis of ATP. The efficiency of this system determines how well the body converts food into usable energy. When mitochondria function optimally, energy levels remain high, supporting physical endurance, mental clarity, and overall vitality. Conversely, mitochondrial dysfunction results in fatigue, decreased performance, and susceptibility to chronic disease.

5.4.2 Mitochondria and Cellular Health

Beyond energy production, mitochondria are deeply involved in maintaining cellular health. They regulate apoptosis, or programmed cell death, which is essential for removing damaged or dysfunctional cells. They also generate reactive oxygen species as byproducts of energy metabolism, which serve as important signaling molecules when maintained in balance. However, excessive oxidative stress damages DNA, proteins, and membranes, impairing both mitochondrial and cellular function. Mitochondria also play a role in calcium regulation, influencing muscle contraction, neurotransmitter release, and intracellular communication. These broader functions highlight that mitochondrial health is not only about energy output but also about maintaining the integrity and balance of the entire cellular environment.

5.4.3 Link between Mitochondria and Aging

Mitochondria are at the center of theories of aging. Over time, mitochondrial DNA accumulates mutations, oxidative stress increases, and the efficiency of energy production declines. These changes contribute to cellular senescence, tissue degeneration, and the decline in physical and cognitive function associated with aging. The mitochondrial free radical theory of aging suggests that damage from reactive oxygen species plays a central role in the aging process. More recent research points to mitochondrial dysfunction as both a cause and consequence of aging, creating a feedback loop that accelerates decline. Supporting mitochondrial health, therefore, becomes a vital strategy for promoting longevity and delaying age-related diseases.

5.4.4 Mitochondrial Dysfunction and Disease

Mitochondrial dysfunction is implicated in a wide range of chronic illnesses. Metabolic disorders such as diabetes and obesity are closely tied to impaired mitochondrial efficiency. Neurodegenerative diseases including Alzheimer's and Parkinson's involve disruptions in mitochondrial function that impair neuronal survival. Cardiovascular disease, chronic fatigue syndrome, and autoimmune conditions also show strong links to mitochondrial health. Because mitochondria are central to energy metabolism and cellular signaling, their dysfunction has widespread consequences. Optimizing mitochondrial health can, therefore, serve not only as a tool for performance enhancement but also as a preventative and therapeutic strategy against chronic disease.

5.4.5 Lifestyle Strategies for Mitochondrial Optimization

Several lifestyle practices have been shown to support mitochondrial health. Exercise remains one of the most effective interventions, as it stimulates mitochondrial biogenesis, the process of creating new mitochondria, and enhances their efficiency. Both endurance and high-intensity training activate pathways that improve mitochondrial density and function. Nutrition also plays a critical role, with diets rich in antioxidants, polyphenols, and healthy fats supporting mitochondrial resilience. Intermittent fasting and caloric restriction activate adaptive stress responses that improve mitochondrial efficiency and stimulate repair. Sleep is equally vital, as mitochondrial maintenance and repair processes are closely tied to circadian rhythms and restorative rest.

5.4.6 Nutritional Support for Mitochondria

Nutrition provides the raw materials and cofactors necessary for mitochondrial function. B vitamins, including B1, B2, B3, and B5, are essential for energy metabolism within the mitochondria. Coenzyme Q10 plays a key role in the electron transport chain, and supplementation has been shown to improve mitochondrial efficiency, particularly in aging individuals. Antioxidants such as vitamin C, vitamin E, and glutathione help buffer reactive oxygen species, preventing oxidative damage. Polyphenols found in foods like green tea, blueberries, and red wine activate protective pathways that support mitochondrial resilience. Omega-3 fatty acids, particularly DHA, are important for mitochondrial membrane health and function. A nutrient-dense diet that emphasizes whole foods supports optimal mitochondrial energy production and defense.

5.4.7 Biohacking Tools for Energy Optimization

Biohackers have explored a wide range of tools and interventions to enhance mitochondrial performance. Red and near-infrared light therapy, also known as photobiomodulation, has been shown to stimulate mitochondrial activity by enhancing the function of cytochrome c oxidase in the electron transport chain. Cold exposure and sauna use introduce hormetic stressors that trigger mitochondrial adaptation and resilience. Nootropics and supplements, including compounds like PQQ, NAD+ precursors, and acetyl-L-carnitine, are used to support mitochondrial biogenesis and energy metabolism. Wearables that track heart rate variability, sleep, and recovery allow individuals to assess the effectiveness of their interventions on mitochondrial health and energy output. These technologies represent a modern layer of mitochondrial biohacking that complements traditional lifestyle strategies.

5.4.8 Role of NAD+ in Energy Optimization

Nicotinamide adenine dinucleotide, or NAD+, is a coenzyme that plays a central role in mitochondrial function and cellular energy metabolism. Levels of NAD+ decline with age, contributing to mitochondrial inefficiency and cellular dysfunction. Boosting NAD+ through precursors such as nicotinamide riboside or nicotinamide mononucleotide has gained attention as a longevity strategy. By enhancing NAD+ availability, these compounds support mitochondrial energy production, DNA repair, and cellular resilience. Research into NAD+ supplementation continues to grow, offering promising avenues for improving both mitochondrial health and overall longevity.

5.4.9 Stress, Hormesis, and Mitochondrial Resilience

Mitochondria respond positively to hormetic stressors, which introduce mild challenges that stimulate adaptive responses. Exercise, fasting, thermal stress, and hypoxia all activate pathways that strengthen mitochondrial efficiency and repair. These stressors upregulate protective proteins, stimulate biogenesis, and enhance the capacity of mitochondria to handle future challenges. By incorporating hormesis into daily routines, individuals can train their mitochondria much as they train muscles, improving resilience and energy output over time. This adaptation-based approach is central to physical biohacking and highlights the dynamic nature of mitochondrial health.

5.5 GUT HEALTH AND MICROBIOME OPTIMIZATION

5.5.1 Foundation of Well-Being: Understanding the Gut Microbiome

The human gut microbiome is a diverse and densely populated metropolis, with its composition unique to each individual, shaped by genetics, mode of birth, early life nutrition, diet, environment, and medication use. These microbes are not passive passengers; they are active participants in human metabolism. They possess enzymes that humans lack, allowing them to break down complex dietary fibers and polyphenols that would otherwise be indigestible, transforming them into essential nutrients and a suite of beneficial compounds.

The most crucial of these microbial metabolites are short-chain fatty acids (SCFAs), such as butyrate, propionate, and acetate. Butyrate, in particular, serves as the primary fuel source for the cells lining the colon, strengthening the gut barrier and preventing a condition often referred to as "leaky gut," where toxins and undigested food particles escape into the bloodstream, triggering inflammation. Beyond local effects, these microbial messengers regulate the immune system, with approximately seventy percent of immune cells residing in the gut-associated lymphoid tissue. They also produce neurotransmitters, including a significant portion of the body's serotonin and gamma-aminobutyric acid (GABA), directly influencing brain function and emotional state through the gut-brain axis. A balanced, diverse microbiome is therefore synonymous with resilience, while a state of dysbiosis—an imbalance in microbial communities—is increasingly linked to a staggering array of conditions, including inflammatory bowel disease, obesity, type 2 diabetes, autoimmune disorders, anxiety, depression, and neurodegenerative diseases.

5.5.2 Nutritional Levers: Feeding the Microbial Community

The most direct and powerful tool for microbiome optimization is diet, as the foods we consume directly select for the growth of specific microbial populations. The cornerstone of a microbiome-supportive diet is dietary fiber, specifically prebiotic fibers that resist digestion in the upper GI tract and serve as fermentable fuel for beneficial bacteria in the colon. Rich sources of prebiotics include chicory root, garlic, onions, leeks, asparagus, Jerusalem artichokes, dandelion greens, and underripe bananas.

The consistent consumption of these foods encourages the proliferation of bacteria that produce beneficial SCFAs, which in turn lower gut pH, creating an environment that discourages the growth of pathogenic species. Beyond prebiotics, a diverse diet rich in a wide variety of plant-based foods is critical. Research suggests that consuming over thirty different plant types per week is associated with a more diverse microbiome, as different fibers and phytonutrients support different microbial niches.

Polyphenols, the antioxidant compounds found in brightly colored berries, green tea, dark chocolate, and extra virgin olive oil, are also not fully absorbed by us but are metabolized by gut microbes into anti-inflammatory compounds. Conversely, a diet high in ultra-processed foods, refined sugars, and unhealthy fats can promote the growth of pro-inflammatory microbes and yeasts, contributing to dysbiosis and weakening the intestinal barrier. Therefore, the primary nutritional biohack for gut health is a shift towards a whole-foods, plant-rich diet, abundant in fiber and phytonutrients.

5.5.3 Strategic Supplementation and Fermented Foods: Reseeding and Reinforcing

While diet lays the foundation, targeted supplementation can address specific imbalances and enhance microbial diversity. Probiotics are live microorganisms that, when administered in adequate amounts, confer a health benefit on the host. They can be found in supplement form or in traditionally fermented foods. However, it is crucial to understand that consuming probiotics is not like planting a permanent garden; these are transient microbes that can confer benefits during their passage, such as crowding out pathogens, producing vitamins like B12 and K2, and modulating immune function, but they typically do not colonize the gut long-term. The selection of a probiotic strain should be evidence-based and targeted to a specific need, such as Lactobacillus rhamnosus GG for travelers' diarrhea or Saccharomyces boulardii for antibiotic-associated issues. Incorporating a wide array of fermented foods into the diet—such as live-culture yogurt, kefir, raw sauerkraut, kimchi, miso, tempeh, and kombucha—provides a diverse array of probiotic strains alongside their own beneficial metabolites. For a more profound and lasting shift, prebiotic supplements like resistant starch (from green banana flour or cooked and cooled potatoes and rice), acacia fiber, or partially hydrolyzed guar gum can be used to selectively nourish desirable resident bacteria already present in the gut, helping them to thrive and outcompete less beneficial species.

5.5.4 Advanced Biohacking and Lifestyle Integration: Beyond Diet

Optimizing gut health extends beyond nutrition into broader lifestyle factors that significantly impact the microbial ecosystem. Chronic stress is a major disruptor of gut harmony, activating the sympathetic nervous system and altering gut motility, secretion, and permeability, while also shifting the microbial composition towards a less favorable profile. Therefore, stress-management practices such as mindfulness meditation, yoga, and breathwork are not merely for mental calm but are essential biohacks for gut health.

Circadian rhythm disruption, through irregular sleep patterns or exposure to artificial light at night, can also negatively alter the gut microbiome's composition and function. Prioritizing consistent, high-quality sleep is thus a non-negotiable component of microbiome optimization. Furthermore, regular physical activity has been shown to increase microbial diversity and enhance the production of beneficial SCFAs, independent of dietary changes. Even environmental factors play a role; exposure to diverse natural environments (the "green" and "blue" spaces) can increase microbial diversity through contact with environmental microbes. For those seeking advanced insights, commercial microbiome testing kits can provide a snapshot of one's microbial composition, offering clues about diversity and the relative abundance of various taxa. While interpretation requires caution and professional guidance, this data can inform personalized dietary and supplement strategies, moving from generalized advice to a more tailored approach for restoring and maintaining a healthy, resilient gut ecosystem, which in turn forms the bedrock of lifelong health and vitality.

CHAPTER 6

Genetic and Molecular Biohacking

6.1 DIY GENETIC TESTING AND ANALYSIS

In recent years, the concept of biohacking has moved beyond nutrition, sleep, and exercise optimization to include genetic and molecular manipulation. A growing community of enthusiasts, researchers, and individuals curious about their biological identity has embraced DIY genetic testing and analysis. This practice involves using accessible technologies such as direct-to-consumer (DTC) DNA kits, open-source bioinformatics tools, and community laboratories to decode and interpret genetic data without relying exclusively on clinical institutions. Often described as genetic and molecular biohacking, this field empowers individuals to explore their genomes, identify health risks, optimize performance, and sometimes even attempt personal experimentation at the molecular level. While it offers immense opportunities, it also raises ethical, privacy, and safety concerns.

6.1.1 Evolution of DIY Genetic Testing

Genetic testing was once confined to advanced medical research labs with expensive sequencing equipment and highly trained professionals. The Human Genome Project, completed in 2003, took more than a decade and billions of dollars to map a single human genome. However, technological innovation has reduced both cost and time dramatically. Today, commercial kits from companies like 23andMe, AncestryDNA, and Nebula Genomics allow consumers to obtain partial or full genome data for a few hundred dollars.

The DIY movement emerged as enthusiasts sought to take this data further, moving beyond ancestry information to investigate deeper aspects of health, disease predispositions, and even traits related to personality or physical performance. The rise of open-source bioinformatics software, cloud-based analysis platforms, and community labs provided the necessary tools to make genetic exploration accessible to non-experts.

6.1.2 Tools and Technologies in DIY Genetic Analysis

DIY genetic testing relies on several tools and approaches, ranging from consumer kits to advanced sequencing technologies.

- **Direct-to-Consumer Kits:** These include saliva-based DNA collection kits that allow individuals to send samples to a laboratory and receive processed genetic data. They typically offer ancestry analysis and health-related insights.
- **Whole Genome Sequencing (WGS):** Some companies now provide full genome sequencing services at reduced costs, giving enthusiasts a complete dataset to work with.
- **Open-Source Software:** Tools such as PLINK, Ensembl Genome Browser, and Galaxy Project enable individuals to analyze raw DNA data, identify genetic markers, and compare variants with reference genomes.
- **Community Biohacking Labs:** Spaces such as DIYBio and local biohacker collectives provide access to equipment like PCR machines, CRISPR kits, and DNA extraction tools for personal experiments.
- **Cloud-Based Platforms:** Online services allow individuals to upload raw DNA files for deeper interpretation of health markers, rare genetic variants, and personalized nutrition or fitness insights.

These resources collectively form the foundation of genetic and molecular biohacking, where individuals attempt to understand and sometimes modify their biological blueprint.

6.1.3 Applications of DIY Genetic Testing and Biohacking

The popularity of DIY genetic testing is not only rooted in curiosity but also in practical applications. Several areas of life and health benefit from such exploration.

- **Health and Disease Risk Analysis:** Many people use DIY testing to identify **predispositions to conditions** such as diabetes, heart disease, Alzheimer's, or certain cancers. While not diagnostic, these results can motivate lifestyle changes and encourage preventive care.
- **Nutrigenomics and Personalized Nutrition:** Genetic variations affect how individuals metabolize nutrients. DIY analysis can reveal sensitivities to caffeine, lactose, or gluten, guiding dietary decisions for improved health.
- **Fitness and Performance Optimization:** Some genetic markers influence muscle composition, oxygen uptake, or recovery speed. Athletes and fitness enthusiasts use this information to tailor training and recovery strategies.
- **Ancestry and Identity Exploration:** For many, genetic testing is an entry point to discovering heritage, uncovering family histories, and connecting with distant relatives.

- **Gene Editing Experiments:** Advanced biohackers experiment with tools like CRISPR-Cas9 for personal projects, including bacterial resistance or simple model organism studies. While controversial, this represents the cutting edge of molecular biohacking.
- **Forensic and Personal Curiosity:** Some enthusiasts analyze DNA to identify traits like eye color, taste preferences, or even behavioral tendencies, blending science with curiosity.

6.1.4 Benefits of DIY Genetic Biohacking

The rise of DIY genetic analysis provides a range of benefits for individuals and society.

- **Empowerment of Individuals:** People gain control over their genetic data, allowing for informed decisions about health and lifestyle.
- **Accessibility of Science:** Complex genetic research is no longer confined to professionals but open to citizen scientists and hobbyists.
- **Early Detection and Prevention:** Awareness of predispositions may lead to early interventions and healthier habits.
- **Innovation and Discovery:** Citizen-driven projects sometimes generate new insights into biology, expanding collective knowledge.
- **Personalization:** Health, diet, and fitness can be customized to genetic tendencies, making optimization more effective.

6.1.5 Risks and Ethical Concerns

Despite its appeal, DIY genetic biohacking is not without challenges. Several risks and ethical dilemmas surround the practice.

- **Privacy Issues:** Genetic data is highly sensitive. Uploading raw DNA files to third-party sites can expose individuals to data breaches, unauthorized research, or misuse by corporations.
- **Misinterpretation of Data:** Without professional guidance, individuals may misunderstand their results, leading to unnecessary anxiety or false reassurance.
- **Lack of Regulation:** Many DIY kits and platforms operate in a regulatory gray area, raising concerns about accuracy and safety.
- **Potential Misuse of Gene Editing:** Biohackers experimenting with CRISPR at home face the risk of unintended mutations or unsafe applications.
- **Ethical Boundaries:** Genetic testing can reveal sensitive information about family relationships, identity, or disease risks, creating dilemmas regarding disclosure and consent.

6.1.6 Case Studies and Examples

A few real-world examples illustrate how DIY genetic testing and biohacking manifest.

- **DIYBio Communities:** Groups around the world have established labs where members share resources for DNA extraction, genetic engineering, and molecular biology experiments.
- **OpenSNP Project**: A platform where individuals can upload and share their raw genetic data openly, fostering collaborative research and data analysis.
- **CRISPR DIY Kits**: Commercially available kits allow hobbyists to experiment with gene editing in bacteria, demonstrating both the accessibility and controversy of biohacking.
- **Athletic Applications**: Some sports enthusiasts have used DNA insights to modify training programs, though governing bodies often caution against misuse in competitive contexts.

6.2 EPIGENETICS AND GENE EXPRESSION MODULATION

6.2.1 Mechanisms of Epigenetic Control: The Molecular Switches

Epigenetic regulation is achieved through several key biochemical processes that act like software instructions telling the DNA hardware what to do. These mechanisms physically modify the DNA or its associated proteins to make genes more or less accessible to the cellular machinery that reads them.

- **DNA Methylation:** This process involves the addition of a methyl group (a carbon and three hydrogens) to a cytosine base in the DNA molecule, typically at sites called CpG islands. This acts primarily as a "dimmer switch," silencing gene expression. Densely methylated regions are typically transcriptionally inactive, making this a crucial process for cellular differentiation (e.g., ensuring a liver cell doesn't start expressing neuron-specific genes) and genomic imprinting.
- **Histone Modification:** DNA is not naked within the nucleus; it is tightly wrapped around proteins called histones, forming a complex known as chromatin. The tails of these histone proteins can be chemically modified in various ways:
  - **Acetylation:** The addition of an acetyl group neutralizes the histone's positive charge, loosening the chromatin structure (euchromatin). This open configuration allows transcription factors to bind more easily, **activating** gene expression.
  - **Methylation:** The effect of histone methylation depends on the specific amino acid methylated. It can either activate or repress transcription. For

example, methylation of histone H3 at lysine 4 (H3K4me) is an activation mark, while methylation at lysine 27 (H3K27me) is a repression mark.

- **Other Modifications:** This includes phosphorylation, ubiquitination, and sumoylation, each adding another layer of fine-tuning to gene expression.

- **Non-Coding RNA (ncRNA) Involvement:** Not all RNA is translated into protein. Small non-coding RNAs, like microRNAs (miRNAs) and small interfering RNAs (siRNAs), play a significant epigenetic role. They can bind to messenger RNA (mRNA) molecules, targeting them for degradation or blocking their translation into protein, effectively **silencing specific genes** post-transcriptionally.

These mechanisms work in concert, creating a complex and responsive epigenetic landscape that dictates cellular identity and function without changing the DNA sequence itself.

6.2.2 The Biohacker's Toolkit: Modulating Your Epigenome

The most exciting aspect of epigenetics is its inherent dynamism and responsiveness to environmental cues. This plasticity is the foundation of molecular biohacking—the deliberate application of interventions to steer epigenetic markers toward a desired health outcome. This is not about changing your genes but about changing how your genes behave.

Nutritional Biohacking (Nutrigenomics)

The famous adage "you are what you eat" finds a profound truth in epigenetics. Specific nutrients directly influence the epigenetic machinery.

- **Methyl Donors:** Compounds like folate (vitamin B9), vitamin B12, choline, and betaine provide the necessary methyl groups for DNA methylation. Adequate intake is crucial for proper methylation patterns, which influence everything from neurotransmitter production to cancer suppression.

- **Polyphenols:** Found in abundance in green tea (EGCG), turmeric (curcumin), berries, and broccoli sprouts (sulforaphane), these compounds can act as histone deacetylase (HDAC) inhibitors. By inhibiting the enzymes that remove acetyl groups, they promote a more open, active chromatin state, potentially activating protective genes like those for antioxidant production and detoxification.

- **Resveratrol:** Found in red grapes and blueberries, resveratrol activates a class of enzymes called sirtuins (SIRT1), which are deacetylases linked to longevity and metabolic health through their role in histone modification and DNA repair.

Lifestyle and Environmental Modulation

Our daily habits exert a powerful and continuous epigenetic influence.

- **Exercise:** Physical activity is a potent epigenetic modulator. It has been shown to alter methylation patterns on genes involved in energy metabolism and inflammation. For instance, exercise can demethylate (activate) the *PPARGC1A* gene, a master regulator of mitochondrial biogenesis, effectively teaching your body to build more energy-producing powerhouses.
- **Stress Management:** Chronic stress leads to the prolonged release of cortisol, which can dysregulate the epigenetic control of genes involved in the immune response and mental health. Practices like mindfulness, meditation, and quality sleep promote a healthy epigenetic profile by reducing cortisol and supporting the body's natural repair cycles.
- **Sleep Hygiene:** Quality sleep is essential for epigenetic maintenance. Disrupted sleep patterns are linked to aberrant methylation of genes regulating circadian rhythms, metabolism, and stress response, increasing the risk of chronic disease.
- **Toxin Avoidance:** Exposure to environmental toxins like cigarette smoke, heavy metals, and endocrine-disrupting chemicals (plastics, pesticides) can cause widespread epigenetic havoc, leading to inappropriate gene silencing or activation. Minimizing exposure is a foundational biohacking strategy.

Technological and Advanced Interventions

While still largely in the experimental realm, cutting-edge technologies point to the future of epigenetic biohacking.

- **Epigenetic Testing:** Companies offer tests that analyze DNA methylation patterns to provide a "biological age" estimate (e.g., Horvath's epigenetic clock) and insights into metabolic and physiological health. Biohackers use this data to track the effectiveness of their interventions over time.
- **Pharmacological Agents:** Certain drugs, known as epi-drugs, are designed to target the epigenetic machinery. For example, DNA methyltransferase and HDAC inhibitors are used in cancer therapy to reactivate silenced tumor suppressor genes. Their use outside of clinical settings is not recommended due to potential side effects but represents a powerful proof-of-concept.
- **CRISPR-Based Epigenetic Editing:** The next frontier involves using modified CRISPR-Cas systems that can target specific epigenetic

modifiers to precise locations in the genome. Instead of cutting the DNA, the Cas enzyme is fused to an enzyme that can add or remove an epigenetic mark (e.g., a methyl or acetyl group). This offers the potential for incredibly precise gene expression modulation, though it remains an experimental tool.

6.3 AT-HOME CRISPR AND GENE MODIFICATION PROJECTS

6.3.1 CRISPR and How Does It Work?

CRISPR (Clustered Regularly Interspaced Short Palindromic Repeats) is a gene-editing technology derived from a bacterial immune system. It allows scientists to precisely modify DNA by cutting, adding, or altering specific genetic sequences.

- **Guide RNA (gRNA):** A short RNA sequence that directs the Cas9 protein to a specific DNA target.
- **Cas9 Protein:** An enzyme that acts as molecular scissors, cutting the DNA at the location specified by the gRNA.

Once the DNA is cut, the cell's natural repair mechanisms can either:

- **Disable a gene** (knockout) by introducing errors during repair.
- **Insert a new gene** (knock-in) by providing a DNA template for repair.

Why CRISPR is Revolutionary

- **Precision:** Unlike older gene-editing methods, CRISPR is highly accurate.
- **Affordability:** CRISPR kits are now available for under $200, making them accessible to hobbyists.
- **Versatility:** It can be used on bacteria, plants, animals, and even human cells.

6.3.2 Tools and Resources for At-Home CRISPR Projects

CRISPR Kits

Several companies offer DIY CRISPR kits designed for educational and experimental purposes:

- **The ODIN CRISPR Kit:** A beginner-friendly kit that allows users to modify bacteria (e.g., *E. coli*) to become antibiotic-resistant.
- **Amino Labs' CRISPR Kit:** Designed for educational use, enabling users to edit genes in yeast or bacteria.
- **Bio-Rad's pGLO Kit:** A simpler alternative that introduces students to genetic engineering by making bacteria glow.

Basic Lab Equipment

To perform CRISPR Experiments at Home, Biohackers Need:

- **PCR Machine (Thermocycler):** For amplifying DNA.
- **Gel Electrophoresis Setup:** To analyze DNA fragments.
- **Micropipettes and Sterile Workspace:** For precise liquid handling.
- **Incubator:** To grow modified organisms under controlled conditions.

Online Communities and Learning Resources

- **Biohacking Forums (e.g., DIYBio, Reddit's r/biohackers):** Where enthusiasts share protocols and troubleshooting tips.
- **Open-Source Protocols:** Websites like Protocol Online and Addgene provide free CRISPR guides.
- **YouTube Tutorials:** Many biohackers document their experiments, offering step-by-step guidance.

6.3.3 Potential Applications of At-Home CRISPR Projects

Modifying Bacteria for Art and Education:

- **Glowing Bacteria:** Using CRISPR to insert bioluminescent genes (e.g., from jellyfish) into E. coli.
- **Antibiotic Resistance Studies:** Editing bacteria to study how resistance develops.

Plant Genetic Engineering:

- **CRISPR-Edited Houseplants:** Modifying plants to change color, size, or resistance to pests.
- **Faster-Growing Crops:** Experimenting with genes that enhance growth rates.

Personalized Medicine and Health Experiments:

- **Gene Therapy Research:** Some biohackers explore CRISPR for potential health benefits, though this is highly controversial.
- **DIY Diagnostics:** Using CRISPR-based tools (like SHERLOCK) to detect diseases at home.

Synthetic Biology and Bioart:

- **Creating Novel Organisms:** Combining genes from different species to produce unique traits.
- **Bioart Projects:** Using genetically modified organisms in artistic installations.

6.3.4 Ethical and Safety Concerns

Biosafety Risks:

- **Unintended Mutations:** Off-target effects could create harmful organisms.
- **Pathogen Creation:** Accidental or intentional modification of dangerous microbes.
- **Environmental Contamination:** Modified organisms could escape and disrupt ecosystems.

Ethical Dilemmas:

- **Human Gene Editing:** Editing human embryos or germ cells raises serious ethical questions.
- **Dual-Use Potential:** CRISPR could be misused for bioterrorism or unethical experiments.
- **Lack of Regulation:** Many countries have no laws governing at-home genetic engineering.

Legal and Regulatory Challenges:

- **FDA and EPA Restrictions:** In the U.S., modifying certain organisms may require approval.
- **International Laws:** Some countries ban CRISPR experiments outside licensed labs.

6.4 DNA DATA STORAGE AND ENCRYPTION

6.4.1 DNA as a Medium for Information Storage

DNA is nature's storage device. Every living organism uses DNA to encode genetic instructions with remarkable efficiency and stability. In theory, one gram of DNA can store nearly 215 petabytes of data, far exceeding the capacity of traditional electronic media. This extraordinary density arises from DNA's four-letter nucleotide alphabet—adenine (A), thymine (T), cytosine (C), and guanine (G)—which can be used to represent binary information. Researchers encode digital files by converting binary data (0s and 1s) into nucleotide sequences (A, T, C, G). Synthetic biology techniques then allow for the physical creation of DNA strands carrying this information. Later, sequencing machines decode the strands back into digital form. This process demonstrates the extraordinary potential of biological molecules as data carriers, offering durability for thousands of years under proper conditions.

6.4.2 Role of Encryption in DNA Data Storage

While DNA storage promises unparalleled density and longevity, it also presents unique security concerns. Data stored in DNA could be read by anyone with access to sequencing technology, making encryption essential. Encryption in DNA data storage involves applying cryptographic methods to secure the encoded sequences, ensuring only authorized parties can access the information. Approaches to encryption in DNA storage include:

- **Classical Cryptography**: Converting binary data into nucleotide sequences after applying standard encryption algorithms such as AES or RSA.
- **Biological Steganography**: Hiding encrypted data within non-coding or randomized DNA sequences to obscure its presence.
- **DNA-Based Cryptography**: Leveraging biological principles, such as complementary base pairing, to develop novel encryption schemes where only a matching sequence can unlock the data.

6.4.3 Molecular Biohacking and DNA Data Experiments

DIY biohackers and community laboratories are increasingly exploring the possibilities of DNA data storage, often experimenting with small-scale encoding and decoding processes. These initiatives may not rival industrial-scale projects, but they demonstrate how molecular biohacking can democratize advanced technologies. Biohackers typically engage in the following:

- **Encoding Simple Messages**: Converting short text files into nucleotide sequences and synthesizing DNA strands to physically store them.
- **DNA Steganography**: Embedding messages within genetic material of bacteria or viruses, a controversial but innovative form of encryption.
- **Synthetic Biology Kits**: Using commercially available CRISPR and DNA synthesis kits to experiment with encoding personalized data at home or in community labs.

6.4.4 Advantages of DNA Data Storage and Encryption

- **Density and Capacity:** DNA's storage density far exceeds any existing electronic medium, offering potential to store all global digital data in a few kilograms of DNA.
- **Longevity:** Unlike magnetic tapes or hard drives that degrade in years or decades, DNA can remain stable for thousands of years if kept in suitable conditions.
- **Energy Efficiency:** Once synthesized, DNA requires minimal energy for storage compared to massive energy costs of data centers.

- **Security Potential:** When combined with encryption and steganographic methods, DNA data storage offers unique pathways for secure, hidden information systems.
- **Biological Integration:** DNA can, in theory, be embedded into living organisms, creating entirely new paradigms of bio-storage and biological computing.

6.4.5 Challenges and Risks

- **Cost:** Synthesizing and sequencing DNA remains expensive compared to electronic storage methods, though costs are steadily declining.
- **Speed:** Writing and reading DNA data is far slower than accessing traditional hard drives or SSDs.
- **Error Rates:** DNA synthesis and sequencing introduce errors such as mutations, insertions, or deletions, requiring robust error-correcting codes.
- **Bioethical Issues:** Embedding data into living organisms or using genetic material as storage raises ethical and safety concerns.
- **Accessibility:** Advanced sequencing tools are not universally available, limiting widespread adoption in DIY or citizen science communities.

6.4.6 Applications in Security and Biohacking

- **National Security:** Governments could archive sensitive intelligence in DNA stored within secure facilities, reducing risks of digital hacking.
- **Medical Records:** Genetic and medical data could be encrypted and stored in DNA to preserve privacy across generations.
- **Digital Time Capsules:** DNA's durability makes it ideal for archiving cultural, historical, and scientific information for millennia.
- **Biohacker Projects:** Enthusiasts experiment with embedding encrypted messages in microbial DNA or using genetic codes for artistic or philosophical statements about data and identity.

6.4.7 Case Studies and Notable Experiments

- **Harvard University (2012):** Researchers encoded a book into DNA, demonstrating the feasibility of large-scale text storage.
- **Microsoft and University of Washington:** Collaborated to build automated DNA storage systems, moving toward practical deployment.
- **DNA Cryptography Studies:** Researchers developed DNA-based one-time pads and other cryptographic methods, showcasing hybrid models of biology and information security.

- **Artistic Biohacking:** Artists and hobbyists have embedded music, poetry, and images into DNA strands as symbolic acts of merging biology with culture.

6.4.8 Ethical and Regulatory Considerations

- **Data Privacy:** How should ownership of DNA-encoded information be managed, especially if embedded in biological material?
- **Dual-Use Concerns:** While DNA encryption could secure sensitive data, it might also enable malicious actors to conceal harmful information or instructions.
- **Environmental Risks:** Embedding DNA data in organisms could inadvertently affect ecosystems if released.
- **Equity and Access:** Ensuring that DNA storage technology does not become monopolized by corporations or governments remains a challenge.

6.5 ETHICAL CONCERNS IN GENETIC SELF-EXPERIMENTATION

6.5.1 Allure and the Ethos: Why Biohackers Experiment on Themselves

- **The Right to Autonomy:** A core tenet is the belief that individuals have an absolute right to explore and alter their own biology. This is framed as a fundamental question of bodily sovereignty and informed consent, rejecting external authority from medical institutions or government agencies.
- **Democratization of Science:** Many biohackers aim to break down the barriers to cutting-edge science, making tools and knowledge available to all, not just those within academia or pharmaceutical companies. They argue that this accelerates innovation and challenges entrenched interests.
- **Personal Optimization:** A significant driver is the "quantified self" movement, extended to the molecular level. The goal is to enhance cognitive function, extend lifespan, improve physical performance, or alleviate personal health conditions that individuals feel have been inadequately addressed by conventional medicine.
- **Curiosity and Exploration:** The simple desire to learn, experiment, and push the boundaries of human knowledge is a powerful motivator, hearkening back to an era of amateur scientists and explorers.

6.5.2 Core Ethical Dilemmas: A Landscape of Risk

The Paramount Issue: Safety and Informed Consent

This is the most immediate and serious ethical concern. Traditional clinical research is governed by strict protocols to minimize risk, including preclinical trials, Institutional Review Board (IRB) approval, and rigorous oversight.

- **Unproven Interventions:** Biohackers often use reagents, tools (like CRISPR-Cas9), and protocols that have not undergone rigorous safety testing. The risk of off-target effects (unintended genetic edits), immune reactions, or ineffective treatments is extremely high.
- **The Illusion of Informed Consent:** While a biohacker may consent to the risk, the question is whether that consent is truly *informed*. Without a deep understanding of molecular biology, genetics, and statistics, an individual cannot fully comprehend the complex, long-term, and potentially catastrophic risks. This is a far cry from the informed consent process in clinical trials, which is designed to ensure comprehension.
- **Lack of Medical Supervision:** In the event of an adverse reaction—an infection, an autoimmune response, or cancer—self-experimenters are often unprepared to handle the medical consequences. They may also be reluctant to seek professional help, delaying critical care.

The Lack of Oversight and Accountability

The DIY biohacking community operates in a regulatory gray area. While some aspects are regulated (e.g., ordering certain pathogens is illegal), self-experimentation on one's own DNA largely falls outside the scope of agencies like the FDA.

- **No Peer Review:** Experiments are not subject to peer review, a cornerstone of scientific progress that helps identify flawed methodologies, biases, and incorrect conclusions. This can lead to the spread of dangerous misinformation.
- **No Quality Control:** The sourcing of genetic material (e.g., plasmids), tools, and chemicals is unregulated. Their purity, concentration, and authenticity cannot be guaranteed, introducing massive variables and risks.
- **Unverified Claims:** Successes and failures are often reported in online forums or YouTube videos rather than in scientific literature. This leads to unverified, sensationalized claims that can be easily misinterpreted and replicated by others without understanding the context or risks.

The Specter of Eugenics and Social Inequality

The ability to genetically "enhance" oneself raises the dark shadow of eugenics and could exacerbate social divides.

- **The Rise of Biological Inequality:** If genetic enhancements become feasible and effective, they will likely be available only to the wealthy, creating a new class of "genetically augmented" individuals with significant advantages in health, cognition, and longevity. This could lead to a biological caste system.
- **Coercion and Social Pressure:** Even if not mandated by law, social pressure to undergo enhancements to remain competitive in education, careers, or sports could become immense, eroding true personal choice.
- **Defining "Normal":** The pursuit of genetic optimization pathologizes normal human variation and could devalue those who are unable or unwilling to be "enhanced," undermining principles of human equality and diversity.

Data Privacy and Misuse

Genetic data is uniquely personal and immutable. Biohackers often sequence their own genomes and share data within communities to collaborate.

- **Lack of Secure Infrastructure:** DIY communities may not have the secure infrastructure to protect this highly sensitive data from breaches, theft, or misuse.
- **Insurance and Discrimination:** While laws like the Genetic Information Nondiscrimination Act (GINA) offer some protection in health insurance and employment, this area remains a minefield. Leaked genetic data could potentially be used by life insurers, employers, or other entities for discriminatory purposes.
- **Unintended Consequences for Relatives:** Since you share your DNA with your biological relatives, your decision to publicize your genetic information also exposes theirs, without their consent.

CHAPTER 7

Environmental and Sensory Biohacking

7.1 LIGHT, SOUND, AND EMF EXPOSURE OPTIMIZATION

We exist in a sea of environmental and sensory information, a constant flux of energy that our nervous system has evolved to interpret and respond to for survival. In the modern world, however, this ancient interplay has been dramatically altered. We are bathed in artificial light at all hours, subjected to a cacophony of anthropogenic noise, and enveloped in an entirely new and pervasive layer of man-made electromagnetic fields (EMF).

The emerging practice of environmental and sensory biohacking recognizes that these factors are not merely background elements but are potent biological signals that directly modulate our physiology, cognitive function, and long-term health. This discipline moves beyond passive exposure to active optimization, seeking to curate our personal environments to align with our biological needs, thereby enhancing sleep, focus, mood, and overall vitality. It is the deliberate and strategic manipulation of our sensory inputs to hack our way to a more optimal state of being.

7.1.1 The Master Regulator: Harnessing the Power of Light

Light is the most powerful environmental cue for the human body, serving as the primary Zeitgeber, or time-giver, for our circadian rhythms. Our central circadian clock, located in the suprachiasmatic nucleus of the brain, synchronizes itself to the external world based on the light information received through the eyes. This master clock then orchestrates the rhythms of every cell and system in the body, from hormone production to metabolism. Biohackers leverage this profound connection through strategic timing, quality, and avoidance of light. The goal is to reinforce a strong, coherent circadian rhythm, which is foundational to virtually every aspect of health. Morning sunlight exposure is a cornerstone practice. Viewing the low solar angle light, rich in blue wavelengths, within the first hour of waking provides a potent signal to the brain that the day has begun. This cues the suppression of the sleep hormone melatonin, promotes cortisol awakening response for energy, and sets the timer for its production later that evening. Conversely, as the sun sets, the diminishing light and change in spectrum should be mirrored indoors. This involves minimizing exposure to blue light, which is abundantly emitted by LEDs, screens, and other artificial sources. The use of blue-light blocking glasses, along with the activation of night-shift modes on devices and the use of warm, dim lighting in the evening, supports the natural rise of melatonin, preparing the body for restorative sleep. Beyond timing, biohackers also consider the quality of full-spectrum light during the day to support mood and vitamin D production, while using tools like red and near-infrared light therapy to target mitochondrial function and reduce inflammation at a cellular level.

7.1.2 Sonic Environment: Curating Sound for Focus and Calm

Sound is a form of vibrational energy that the brain is perpetually processing, whether we are consciously aware of it or not. Chronic exposure to unwanted noise, or noise pollution, is a significant biological stressor. It activates the sympathetic nervous system, elevates cortisol levels, disrupts cognitive function, and fragments sleep, even if we do not fully wake.

The biohacking approach to sound involves a two-pronged strategy: eliminating harmful noise pollution and introducing beneficial soundscapes. The first step is to create quiet sanctuaries, particularly in the bedroom and workspace. This can be achieved through sound-dampening materials, high-quality earplugs for sleep, or noise-cancelling headphones to create a bubble of focus in a chaotic environment. The second step is the intentional use of sound for enhancement.

Binaural beats are a popular auditory technology where two slightly different frequencies are played in each ear. The brain perceives a third, phantom frequency—the difference between the two—which can entrain brainwaves into specific states. For example, delta and theta waves are associated with deep sleep and meditation, while alpha and beta waves can support focus and alertness. Similarly, isochronic tones and curated soundscapes featuring natural sounds like flowing water, wind, or birdsong can mask disruptive noises and promote a state of calm concentration or relaxation. The optimization of our sonic environment is about taking control of this auditory input, transforming it from a source of stress into a tool for cognitive and emotional regulation.

7.1.3 Navigating the Electromagnetic Landscape: A Prudent Approach

The topic of electromagnetic fields (EMF) from wireless technologies like Wi-Fi, Bluetooth, and cellular networks is perhaps the most controversial yet rapidly growing area of environmental biohacking. While the non-ionizing radiation emitted by these devices is deemed safe at legal exposure limits by regulatory bodies, a growing number of individuals report symptoms of electromagnetic hypersensitivity, and a body of emerging research suggests potential for biological effects, particularly with long-term, chronic exposure. The precautionary principle guides the biohacker's approach: in the absence of scientific consensus, taking simple, prudent steps to minimize unnecessary exposure is a low-risk, potentially high-reward strategy. This is not about reverting to a pre-technological age but about adopting habits that create distance and reduce dose.

Key practices include turning off Wi-Fi routers at night, keeping phones and other wireless devices out of the bedroom or at least away from the bed, and favoring airplane mode whenever connectivity is not essential. The use of wired accessories, such as ethernet cables for internet, wired headphones, and wired keyboards, significantly reduces close-proximity EMF exposure from Bluetooth and cellular transmissions. For those seeking more advanced mitigation, solutions like shielding materials can be used to create lower-EMF zones within the home. The core philosophy is to be mindful of the cumulative exposure and to make conscious choices that reduce the body's constant defensive burden against this novel environmental stressor, thereby potentially supporting better sleep, reduced inflammation, and improved cellular energy.

7.1.4 Synergistic Integration for Holistic Optimization

The true power of environmental biohacking lies not in isolating each factor but in understanding their profound synergy. A disrupted circadian rhythm from poor light exposure can make an individual more susceptible to the negative effects of noise and EMF stress. Conversely, chronic stress from a disruptive sonic and electromagnetic environment can dysregulate circadian biology, creating a vicious cycle. The optimized environment is therefore a harmonized one. It begins with a morning routine of natural light and movement, setting a strong circadian tone. The work environment is then curated for focus: positioned for access to natural light, protected from intrusive noise with headphones or quiet zones, and distanced from the densest sources of EMF. The evening wind-down is a sacred ritual: dim, warm lights replace harsh overhead lighting, digital devices are retired or filtered, wireless transmitters are switched off, and the home is filled with calming sounds or purposeful silence. The bedroom becomes a temple for recovery—pitch dark, quiet, cool, and with minimal electromagnetic fields. This holistic curation of our sensory environment is a continuous process of self-experimentation. By mindfully shaping the light, sound, and energy fields we inhabit, we move from being passive recipients of our environment to active architects of our own biology, crafting the ideal conditions for our bodies and minds to not just function, but to truly thrive.

7.2 BLUE LIGHT BLOCKING AND CIRCADIAN RHYTHM SYNCHRONIZATION

7.2.1 Role of Blue Light in Human Biology

Blue light, characterized by short wavelengths between 400 and 490 nanometers, has profound biological effects because it interacts with intrinsically photosensitive retinal ganglion cells in the eye. These cells communicate directly with the suprachiasmatic nucleus of the brain, which serves as the central circadian clock. Exposure to blue light suppresses the release of melatonin, the hormone responsible for initiating sleep, while simultaneously increasing cortisol and alertness levels. In natural conditions, this mechanism helps synchronize biological rhythms with the day-night cycle. However, artificial environments filled with electronic screens, LED lighting, and prolonged nighttime illumination extend exposure to blue light far beyond evolutionary norms. This creates a mismatch between the body's internal timing systems and external conditions, a phenomenon often described as circadian misalignment.

7.2.2 Circadian Rhythm and Its Importance

Circadian rhythm is the intrinsic 24-hour cycle that governs physiological, behavioral, and metabolic processes in humans. It regulates sleep-wake patterns, hormone secretion, digestion, and cognitive performance. When circadian rhythms are synchronized with natural environmental cues such as sunrise and sunset, individuals experience restorative sleep, efficient energy use, and balanced hormonal cycles. Conversely, disruptions in circadian timing are linked to insomnia, mood disorders, obesity, diabetes, cardiovascular disease, and even neurodegenerative conditions. In modern society, circadian disruption is widespread due to artificial light exposure at night, irregular sleep schedules, and shift work. For biohackers, restoring circadian alignment through environmental control and sensory intervention is a powerful strategy for optimizing health and performance.

7.2.3 Blue Light Blocking as a Biohacking Strategy

Blue light blocking represents an intentional modification of sensory input to restore natural circadian alignment. It typically involves filtering or reducing exposure to short-wavelength light during evening hours when melatonin production should naturally increase. Specially designed glasses with amber or red-tinted lenses block or attenuate blue wavelengths, allowing the brain to perceive a transition toward darkness. This reduces melatonin suppression and facilitates smoother sleep onset. In addition, many individuals adjust their environment by using warm-toned lighting in the evening, installing software on digital devices that shifts screen colors toward red hues after sunset, or adopting strict screen curfews. Such interventions reflect the biohacker's principle of controlling external stimuli to influence internal biology in measurable and beneficial ways.

7.2.4 Mechanisms of Synchronization

Circadian rhythm synchronization through light management relies on the interaction between environmental light cues and the body's internal clock. Morning exposure to natural blue-enriched daylight acts as a time cue that strengthens circadian alignment, while evening reduction of blue light prevents unwanted delays in sleep onset. By strategically managing exposure, individuals can reinforce natural patterns of hormone release, core body temperature fluctuations, and alertness cycles. The synchronization process works by supporting melatonin secretion at night and enhancing cortisol awakening response in the morning, thereby creating a sharper distinction between daytime activity and nighttime rest. For shift workers or travelers experiencing jet lag, deliberate manipulation of light exposure becomes a practical tool for adjusting the circadian phase more rapidly.

7.2.5 Scientific Evidence and Research

Numerous studies support the effectiveness of blue light blocking in promoting circadian health. Research has shown that wearing blue light blocking glasses in the evening increases melatonin levels, shortens sleep latency, and improves subjective sleep quality. Experimental studies in shift workers demonstrate that minimizing nighttime blue light exposure can reduce circadian misalignment and improve mood and alertness. Clinical investigations also link controlled light exposure to better outcomes in patients with insomnia, seasonal affective disorder, and delayed sleep phase syndrome. Beyond sleep, emerging research suggests that circadian synchronization contributes to metabolic regulation, cardiovascular health, and cognitive resilience, highlighting the far-reaching impact of environmental light control.

7.2.6 Environmental Biohacking Applications

Blue light blocking fits into the broader category of environmental biohacking, which involves manipulating physical surroundings to enhance health and performance. By transforming light environments, individuals reclaim alignment with evolutionary lighting cycles. Homes and workspaces can be outfitted with dynamic lighting systems that shift color temperature throughout the day, mimicking the natural progression of sunlight. Evening environments can be designed to minimize bright, short-wavelength light while maximizing soothing, longer wavelengths that support relaxation. Such approaches extend beyond individual practices to architectural and urban design, where circadian-friendly lighting systems may one day replace static artificial illumination. The underlying principle is to harness environmental cues as allies rather than disruptors of biological rhythms.

7.2.7 Sensory Biohacking and Light Perception

Sensory biohacking emphasizes control over how the body perceives and processes external stimuli. Blue light blocking is a prime example, as it alters the sensory input of vision to guide biological processes. The use of tinted lenses, adaptive screens, and wearable devices represents an attempt to consciously intervene in the interaction between light perception and circadian biology. Unlike pharmacological approaches, sensory biohacking leverages the body's natural mechanisms rather than overriding them. This approach is particularly appealing because it is non-invasive, customizable, and adaptable to personal needs and contexts. Through careful experimentation, individuals can determine the optimal level of blue light reduction that balances evening relaxation with daytime alertness, creating personalized circadian optimization strategies.

7.2.8 Benefits of Blue Light Blocking and Synchronization

The benefits of circadian synchronization through blue light blocking are multifaceted. Improved sleep quality leads to better memory consolidation, emotional regulation, and physical recovery. Enhanced circadian alignment reduces risks of chronic diseases associated with circadian disruption, including obesity and cardiovascular disorders. Daytime productivity and cognitive performance are also heightened when the body is aligned with natural rhythms.

On a psychological level, individuals often report reduced anxiety and stress when they adhere to consistent light-dark cycles. The cumulative effect is not only better health but also enhanced performance and resilience, which explains why biohackers prioritize circadian strategies as foundational to optimization.

7.3 SMART HOME INTEGRATION FOR HEALTH OPTIMIZATION

7.3.1 Role of Smart Homes in Health Optimization

Smart homes are no longer just about automation; they are becoming proactive health partners. By continuously monitoring environmental factors and adjusting conditions in real time, these systems can help prevent illness, improve sleep, boost productivity, and enhance overall well-being. The integration of Internet of Things (IoT) devices allows for seamless communication between different sensors and appliances, creating a cohesive ecosystem that responds to the needs of its inhabitants. For instance, smart thermostats can regulate temperature to support metabolic health, while air purifiers can eliminate pollutants that trigger allergies or respiratory issues. Similarly, smart lighting systems can mimic natural sunlight to regulate circadian rhythms, improving sleep quality and mood. These technologies collectively contribute to a holistic approach to health optimization, where the home itself becomes a therapeutic environment.

7.3.2 Air Quality and Respiratory Health

One of the most critical aspects of a health-optimized smart home is air quality management. Poor indoor air quality has been linked to a range of health problems, including asthma, allergies, and even cognitive decline. Smart air purifiers equipped with HEPA filters and real-time monitoring sensors can detect and remove particulate matter, volatile organic compounds (VOCs), and other harmful pollutants.

Advanced systems can also integrate with smart HVAC units to ensure proper ventilation, reducing the buildup of carbon dioxide and other indoor contaminants. Some devices even use AI to learn household patterns, adjusting filtration levels based on occupancy and outdoor pollution data. For individuals with respiratory conditions, these systems can be lifesaving, providing a consistently clean and safe breathing environment.

7.3.3 Circadian Lighting for Sleep and Mental Health

Light plays a crucial role in regulating the body's internal clock, known as the circadian rhythm. Disruptions in this natural cycle can lead to sleep disorders, mood imbalances, and decreased cognitive function. Smart lighting systems address this by dynamically adjusting color temperature and brightness throughout the day to mimic natural sunlight.

In the morning, these lights can emit cool, blue-enriched light to promote alertness and energy. As evening approaches, they shift to warmer, amber tones to signal the body to produce melatonin, the hormone responsible for sleep. Some systems even sync with wearable devices to personalize lighting based on individual sleep patterns. By optimizing light exposure, smart homes can significantly improve sleep quality, reduce stress, and enhance overall mental well-being.

7.3.4 Temperature Regulation for Metabolic and Cognitive Benefits

Temperature is another environmental factor that profoundly impacts health. Research has shown that slight variations in ambient temperature can influence metabolism, immune function, and even cognitive performance. Smart thermostats, such as those with AI-driven learning capabilities, can maintain optimal temperatures tailored to individual preferences and health goals.

For example, cooler temperatures during sleep have been shown to improve sleep quality and support fat metabolism, while slightly warmer environments may enhance relaxation and recovery. Some advanced systems can even create microclimates within different rooms, ensuring that each space is optimized for its intended use—whether it's a cool home office for focus or a warm bedroom for rest.

7.3.5 Noise Control and Acoustic Biohacking

Soundscapes within a home can significantly affect stress levels, concentration, and overall mental health. Smart home systems can integrate noise-canceling technologies, white noise machines, and even adaptive sound systems to create an acoustically balanced environment. For instance, smart speakers with ambient sound features can generate calming nature sounds or binaural beats to enhance relaxation and meditation.

Additionally, some systems use AI to analyze noise pollution levels and automatically adjust soundproofing measures or play masking sounds to reduce disturbances. This is particularly beneficial in urban environments where external noise can disrupt sleep and increase stress. By optimizing acoustic conditions, smart homes can foster a more peaceful and productive living space.

7.3.6 Water Quality and Hydration Optimization

Water quality is often overlooked in health optimization, yet it is a fundamental aspect of well-being. Smart water filtration systems can monitor and purify tap water, removing contaminants such as heavy metals, chlorine, and microorganisms. Some advanced filters even adjust mineral content to enhance hydration and support metabolic functions.

Furthermore, smart faucets and hydration tracking systems can remind users to drink water at optimal intervals, ensuring proper hydration throughout the day. For those with specific health concerns, such as kidney function or electrolyte balance, these systems can be programmed to deliver water with customized mineral profiles, providing an additional layer of personalized health support.

7.3.7 Role of AI and Machine Learning in Adaptive Homes

Artificial intelligence is the backbone of modern smart home health optimization. Machine learning algorithms analyze data from various sensors—such as air quality monitors, sleep trackers, and wearable devices—to create highly personalized living environments. Over time, these systems learn individual preferences and health patterns, making automatic adjustments to improve well-being.

For example, an AI-driven smart home might detect that a resident is experiencing poor sleep due to high humidity levels and adjust the dehumidifier accordingly. It could also correlate air quality data with respiratory symptoms and suggest interventions before issues arise. The more data these systems collect, the more precise and effective they become in supporting health optimization.

7.3.8 Ethical and Privacy Considerations

While the benefits of smart home health optimization are substantial, they also raise important ethical and privacy concerns. The continuous collection of personal health data, such as sleep patterns, respiratory rates, and even mood indicators, necessitates robust security measures to prevent unauthorized access. Additionally, there is the risk of over-reliance on technology, where individuals may become dependent on automated systems rather than developing natural resilience. There is also the question of accessibility—advanced smart home health technologies may only be available to those with significant financial resources, potentially widening health disparities. As these systems become more integrated into daily life, it will be crucial to establish ethical guidelines and regulations to ensure they are used responsibly and equitably.

7.4 NATURE THERAPY AND GROUNDING TECHNIQUES

7.4.1 The Concept of Nature Therapy

Nature therapy, often referred to as ecotherapy or green therapy, is the practice of spending intentional time in natural environments to support psychological, emotional, and physiological health. It is rooted in the understanding that humans evolved in close interaction with forests, rivers, mountains, and landscapes, and that separation from such environments creates dissonance in mental and physical states. From Japanese forest bathing, known as shinrin-yoku, to contemporary wellness retreats that emphasize time outdoors, the principle is that natural settings reduce stress hormones, enhance immune function, and improve mood. Within the framework of environmental biohacking, nature therapy is regarded as a way to manipulate surroundings deliberately to trigger biological responses that promote health. Instead of relying solely on technological interventions, this approach leverages the restorative power of the natural world to reset human rhythms and enhance resilience.

7.4.2 Sensory Engagement with Nature

The effectiveness of nature therapy lies in the multisensory engagement that natural environments provide. The sound of flowing water, the scent of soil and plants, the visual complexity of green landscapes, and the tactile sensation of natural textures all stimulate the senses in ways that urban environments rarely achieve. Sensory biohacking focuses on how manipulating sensory input can regulate mood, cognition, and physiology, and nature provides a rich palette of sensory experiences to achieve these outcomes. Visual exposure to greenery has been linked to reduced mental fatigue and improved focus, while auditory immersion in birdsong or rustling leaves has been shown to lower blood pressure and promote relaxation. The olfactory input of phytoncides, organic compounds released by plants, strengthens immune function. By immersing the senses in these natural cues, individuals can recalibrate their stress responses and synchronize their internal states with the rhythm of the environment.

7.4.3 Grounding Techniques and the Science of Earthing

Grounding, also known as earthing, is the practice of making direct physical contact with the earth's surface, typically through walking barefoot on grass, sand, or soil, or by using conductive systems that connect the body to the ground. The underlying concept is that the earth carries a natural negative electrical charge, and physical contact allows for the transfer of electrons into the human body. Advocates argue that this exchange reduces inflammation, improves sleep, balances cortisol levels, and stabilizes circadian rhythms. Scientific studies investigating grounding have reported measurable physiological effects, such as improved heart rate variability, reduced pain perception, and enhanced recovery from stress. From a biohacking perspective, grounding is seen as a low-tech but highly effective method of altering biological processes by reestablishing an electrical connection between the body and the natural environment.

7.4.4 Circadian Alignment and Environmental Rhythms

Both nature therapy and grounding contribute to circadian rhythm synchronization, which is central to health and performance. Spending time outdoors exposes individuals to natural light cycles, which are critical for regulating melatonin production and maintaining consistent sleep-wake patterns. The grounding effect further reinforces this alignment by lowering stress hormone levels and calming the nervous system, thereby enhancing the body's ability to transition smoothly between states of alertness and rest. Unlike artificial interventions that attempt to override biological rhythms, these practices restore natural alignment by enhancing the body's sensitivity to environmental cues. This makes them powerful tools for sensory and environmental biohacking, particularly for those struggling with insomnia, jet lag, or stress-related disorders.

7.4.5 Psychological Benefits of Nature Therapy and Grounding

One of the most significant impacts of reconnecting with nature is the improvement in mental health. Research consistently shows that time spent in green spaces reduces symptoms of anxiety, depression, and mental fatigue. Nature therapy provides a sense of calm and perspective that urban settings rarely offer, fostering mindfulness and emotional regulation. Grounding techniques complement this effect by reducing physiological arousal associated with stress and creating a sense of balance and presence. Many practitioners describe grounding as a deeply calming experience that counters the overstimulation of digital environments. For biohackers, these psychological benefits are not abstract but measurable outcomes that contribute directly to improved productivity, creativity, and resilience in daily life.

7.4.6 Immune Function and Physiological Health

Exposure to natural environments and direct grounding practices also have tangible effects on physical health. Forest bathing, for instance, has been linked to increased activity of natural killer cells, a vital component of the immune system.

The anti-inflammatory potential of grounding further supports recovery from chronic pain, injuries, and stress-related immune suppression. Regular practice of nature therapy and grounding has been associated with lower blood pressure, improved circulation, and enhanced energy levels. By integrating these practices into biohacking routines, individuals not only optimize mental clarity but also build resilience at the cellular and systemic levels. These benefits illustrate how environmental and sensory interventions can rival or complement pharmacological approaches to health optimization.

7.4.7 Modern Applications and Accessibility

In the context of modern lifestyles, nature therapy and grounding techniques are being adapted into accessible practices for both urban and rural populations. Guided outdoor sessions, wellness retreats, and corporate initiatives encourage employees to spend time in natural surroundings as part of stress reduction programs. In urban settings, rooftop gardens, indoor plants, and green architecture aim to replicate some of the sensory benefits of nature.

For grounding, specialized conductive mats and footwear provide alternatives for individuals unable to access natural surfaces regularly. While these adaptations may not fully replicate direct natural experiences, they demonstrate the growing recognition of the importance of environmental alignment in health. From a biohacking perspective, these tools extend the reach of nature-based interventions into daily routines and expand their accessibility across diverse lifestyles.

7.5 SENSORY SUBSTITUTION AND AUGMENTATION DEVICES

7.5.1 Principle of Sensory Substitution: Seeing with the Skin and Hearing with the Tongue

The foundational concept behind sensory substitution is the brain's remarkable capacity for neuroplasticity—its ability to reinterpret and repurpose sensory data. These devices function by capturing environmental information typically inaccessible to a user, such as visual imagery or ultrasonic sound, and converting it into a different sensory modality, such as touch or sound, that the user can perceive. The classic example is the tactile-vision sensory substitution system. A camera captures a visual scene, and the resulting image is translated into a grid of vibrating actuators or electrotactile stimulators placed on the user's skin, often on the back, abdomen, or forehead. Initially, the user feels only a pattern of vibrations. However, after dedicated training, a profound cognitive shift occurs. The brain ceases to interpret the signals as mere touch and begins to perceive them as spatial information, allowing the user to navigate environments, recognize objects, and even perceive depth and perspective through their skin. The sensation transitions from a symbolic code to a direct perceptual experience, a testament to the brain's role as an interpreter of patterns rather than a passive receiver of fixed sensory data. Similar systems have been developed for auditory substitution, converting sound into tactile stimuli for the deaf community, and even for conveying infrared or ultraviolet light information. The key insight is that the brain does not care through which channel the data arrives; it can learn to extract meaning from any consistent stream of information.

This process is not without its challenges. It requires significant user training and cognitive load, and the resolution of information conveyed through the skin is inherently lower than that of the biological eye or ear. Nevertheless, these devices offer a powerful alternative for those with sensory impairments and serve as a crucial proof-of-concept for the broader project of sensory augmentation. They demonstrate that our perceptual reality is not absolute but is rather a construct built by the brain from available inputs, and those inputs can be deliberately engineered.

7.5.2 Leap to Sensory Augmentation: Perceiving the Imperceptible

While substitution replaces one sense with another, sensory augmentation aims to add entirely new perceptual capacities to the human experience. These devices expand our senses into domains beyond natural human capabilities, creating what are often termed "superhuman" senses.

A prominent example is the development of magnetic sense. Through a small magnet implanted in the fingertip, a biohacker can gain a constant perception of magnetic fields. The implant vibrates subtly in the presence of electromagnetic sources, allowing the user to feel the presence of live wires, sense the operation of household appliances, or even perceive directionality based on the Earth's magnetic field, much like a built-in compass. This new sense becomes integrated into the user's subconscious perceptual model of the world; they do not need to consciously think about the vibration but simply "know" the magnetic landscape around them.

Other augmentation ventures include echolocation devices, where head-mounted sensors emit ultrasonic clicks and translate the returning echoes into auditory or tactile signals, granting a form of spatial awareness akin to that of bats or dolphins. Some researchers are exploring technologies to perceive real-time data streams, such as stock market fluctuations, weather patterns, or even social media activity, through subtle sensory cues, effectively turning abstract digital information into a constant ambient sense. The goal is to make this information perceptually intuitive, moving it from the realm of conscious intellectual effort to that of effortless, embodied knowing. This represents the ultimate form of environmental biohacking: not just curating existing stimuli but weaving new threads of data directly into the fabric of our sensory reality, allowing us to interact with and understand complex systems in a more immediate and visceral way.

7.5.3 Neurological and Philosophical Implications: Rewiring the Self

The adoption of sensory substitution and augmentation devices carries profound implications that extend far beyond the technical. On a neurological level, long-term use of these technologies literally rewires the brain. Functional MRI studies have shown that using a visual-to-tactile substitution device leads to activation of the visual cortex in response to tactile stimuli. The brain's map is being redrawn, incorporating the new sensory stream into its processing hierarchy. This raises fascinating questions about the nature of perception and consciousness. If a person who has been blind from birth learns to "see" through a camera on their tongue, what is the qualitative nature of their experience? Is it vision, touch, or something entirely new—a unique perceptual modality forged by technology? The line between the senses becomes blurred, challenging our categorical understanding of how we experience the world.

Philosophically, these technologies force a re-examination of the human body's relationship with its tools. A cane used by a blind person is often perceived as an extension of their bodily senses, mapping the world through touch. Sensory augmentation devices represent a radical intensification of this principle, moving from a tool that extends the body to an implant or wearable that becomes integrated into the body's sensory schema. This challenges the very definition of the self and the boundary between the organic body and the technological artifact. It suggests a future of fluid embodiment, where identity is not locked within the biological skin but can be extended and reshaped through integrated cybernetic systems. Furthermore, it introduces the potential for new forms of inequality—a perceptual divide between those with default biological senses and those who can afford to augment their perception, potentially creating classes of individuals who experience and understand reality in fundamentally different ways.

CHAPTER 8

Biohacking Communities and Culture

8.1 GLOBAL BIOHACKING HUBS AND LABS

8.1.1 Emergence of Community Biohacking Spaces

The origins of global biohacking hubs can be traced back to the early 2000s, when the do-it-yourself biology movement began to gain traction. As sequencing costs fell and laboratory equipment became more affordable, citizen scientists sought to create shared spaces where they could experiment without the barriers of academic funding or corporate oversight. These early initiatives gave birth to community laboratories that combined the accessibility of makerspaces with the rigor of scientific experimentation. Over time, these labs expanded into structured organizations offering workshops, mentorship, and collaborative research projects. Today, community biohacking hubs exist in major cities across North America, Europe, Asia, and beyond, making scientific exploration accessible to a wider public than ever before.

8.1.2 Infrastructure and Accessibility of Labs

Global biohacking hubs are often designed to bridge the gap between curiosity and professional research. They typically feature essential laboratory infrastructure such as polymerase chain reaction machines, DNA sequencing tools, cell culture facilities, and bioinformatics software. By providing access to such resources, they enable individuals without institutional affiliation to engage in genetic testing, synthetic biology, and molecular experimentation. Accessibility is not limited to equipment alone but also includes the social infrastructure of shared learning. Many hubs operate as membership-based organizations where knowledge exchange, skill development, and collaborative discovery form the foundation of activity. These laboratories allow biohackers to pursue personal projects, develop start-ups, or contribute to open-source scientific initiatives, all within an inclusive community model.

8.1.3 Notable Global Biohacking Hubs

Several hubs have achieved international recognition for their contributions to the biohacking movement. In the United States, organizations such as Genspace in New York City and Counter Culture Labs in Oakland have pioneered community biology by offering training in molecular techniques and encouraging citizen-led projects. In Europe, La Paillasse in Paris has become a landmark hub combining open science, art, and community engagement, while DIYBio communities in the United Kingdom, Germany, and the Netherlands are creating vibrant networks of experimentation. Asia has also witnessed the rise of significant initiatives, with community biology spaces in Singapore, India, and Japan bringing local perspectives into the global biohacking conversation. These hubs collectively form an interconnected ecosystem where ideas, resources, and practices circulate across borders, expanding the possibilities of nontraditional science.

8.1.4 Collaboration and Knowledge Sharing

A defining characteristic of global biohacking hubs is their commitment to openness and collaboration. Unlike traditional laboratories, which often operate under competitive or proprietary frameworks, community labs emphasize transparency, shared resources, and collective progress. Many hubs organize workshops, seminars, and conferences that bring together biohackers, scientists, artists, and entrepreneurs. International networks such as DIYBio.org have emerged to connect these communities, facilitating the exchange of ideas and ensuring that knowledge generated in one part of the world can inspire experimentation elsewhere. This emphasis on open science reflects the belief that biological knowledge should not be restricted to institutions but should instead empower individuals and communities worldwide.

8.1.5 Role of Technology in Biohacking Labs

Technological innovation is central to the success of global biohacking hubs. Advances in DNA sequencing, genetic engineering, and bioinformatics have lowered the threshold for experimentation, while affordable hardware and software tools have expanded possibilities for citizen science. Open-source equipment, such as low-cost centrifuges or portable PCR machines, makes complex procedures accessible to non-specialists. Software platforms allow individuals to analyze genetic data, simulate biological processes, or design synthetic organisms. Many labs also integrate digital collaboration tools, enabling cross-border projects and fostering a sense of global scientific citizenship. Technology thus transforms community labs into nodes of innovation where experimentation is not only possible but scalable.

8.1.6 Biohacking Hubs as Entrepreneurial Incubators

Beyond being spaces of experimentation, global biohacking hubs increasingly serve as incubators for entrepreneurial initiatives. Start-ups focused on biotechnology, personalized medicine, and environmental sustainability often emerge from community labs. The collaborative atmosphere fosters innovation, and the availability of infrastructure reduces barriers to prototyping and testing. Many hubs host hackathons and innovation challenges that attract participants from diverse fields, sparking interdisciplinary solutions.

This entrepreneurial dimension reflects how biohacking hubs extend beyond hobbyist science to play an active role in shaping the future of biotechnology and human enhancement industries. By blending creativity with practicality, these hubs nurture ventures that may eventually influence mainstream healthcare, agriculture, and environmental management.

8.1.7 Educational and Social Impact

One of the most transformative contributions of global biohacking labs lies in education. These spaces provide hands-on training for individuals who might not otherwise have access to laboratory experience, including students, artists, and enthusiasts from non-scientific backgrounds. By lowering barriers to entry, they inspire new generations to engage with biology not as a distant discipline but as a participatory practice. The social impact of these hubs extends further, as they foster inclusivity, community engagement, and public dialogue about the ethical and societal implications of biotechnology. By encouraging diverse participation, biohacking hubs challenge the exclusivity of traditional science and promote a more democratic vision of innovation.

8.1.8 Ethical Considerations and Challenges

Despite their promise, global biohacking hubs face ethical and regulatory challenges. Safety is a central concern, as community laboratories must ensure that experimentation adheres to biosafety standards and does not pose risks to participants or the public. Regulatory frameworks vary widely across countries, and navigating these differences can be difficult for grassroots initiatives. Ethical dilemmas also arise from the potential misuse of genetic technologies or the blurred boundaries between personal experimentation and clinical application. Community labs often address these challenges by adopting codes of conduct, promoting transparency, and engaging in ongoing dialogue with policymakers and the public. Striking a balance between accessibility and responsibility remains one of the most pressing issues for the biohacking movement.

8.2 BIOHACKER MEETUPS, CONFERENCES, AND NETWORKS

8.2.1 Local Hub: Biohacker Meetups and Community Labs

At the most foundational level, biohacker meetups and community laboratories, often called "hacklabs" or "biofoundries," serve as the local hubs for practical engagement. These spaces, found in major cities worldwide, are democratizing access to the tools of biotechnology. For a monthly membership fee, individuals gain access to centrifuges, PCR machines, gel electrophoresis equipment, and even gene sequencers—instrumentation that would otherwise be prohibitively expensive for an independent citizen scientist. The value of these spaces, however, extends far beyond equipment rental. They are immersive environments for hands-on learning where a software engineer can learn sterile technique from a retired microbiologist, and a college student can get feedback on a DIY CRISPR project from a seasoned bioartist. Regular meetups hosted in these labs or in local cafes cover topics ranging from introductory genetic engineering and mycology to the intricacies of epigenetic testing and nootropic stacks.

The social dynamic of these gatherings is critical. They foster a culture of "show and tell" where individuals present their self-experimentation data, successful or failed, for group analysis. This process provides a form of peer review, where methodologies are questioned, results are scrutinized, and suggestions for improvement are offered in a spirit of collective problem-solving. This collaborative troubleshooting is a vital safety mechanism, potentially identifying flawed protocols or dangerous assumptions before they lead to personal harm. Furthermore, these local networks build trust and accountability. Knowing one's peers face-to-face creates a sense of responsibility and encourages a more rigorous and ethical approach than what might occur in the anonymity of purely online forums. The community lab becomes a testing ground not only for experiments but for the very norms and ethics of the biohacking community itself, establishing best practices through shared experience rather than top-down regulation.

8.2.2 Grand Confluence: Biohacking Conferences and Symposia

If local meetups are the neurons of the movement, then major conferences are its grand synapses—annual or biannual events where the entire community converges to share its most advanced findings and define its future trajectory. Events like the Biohacker Summit in Europe, the Health Optimization Summit, and the Global Biohackers Meetup are more than just conferences; they are cultural phenomena that blend scientific presentation with the energy of a festival. These gatherings feature a diverse roster of speakers, including PhD scientists, visionary entrepreneurs, dedicated self-experimenters, and even performance artists, all united by a common belief in the power of personal biological optimization. The presentations cover the entire spectrum of biohacking, from technical talks on the latest gene editing techniques and neurofeedback protocols to philosophical discussions on the ethics of human enhancement and the future of evolution.

The content delivered from the stage is only one part of the value. The true magic occurs in the hallways, during coffee breaks, and at evening social events. These are the moments where a developer working on a new sensory augmentation device can connect with a materials scientist who can solve a biocompatibility issue. It is where someone struggling to interpret their microbiome sequencing data can find the expert who guides them. These serendipitous collisions of knowledge and need are the catalyst for countless collaborations and projects. Conferences also serve as a powerful legitimizing force for the movement. By attracting mainstream scientists, doctors, and investors, these events demonstrate the growing sophistication and credibility of biohacking, facilitating a necessary dialogue between the establishment and the avant-garde. They are a showcase of the movement's maturity, proving that citizen science can generate valuable insights and innovate at the frontiers of biology.

8.2.3 Digital Cortex: Online Networks and Forums

Underpinning the physical gatherings is a perpetual, global conversation happening in digital spaces. Online platforms constitute the movement's digital cortex, a always-active repository of knowledge and a constant source of support. Dedicated forums on sites like Reddit, Discord servers, and specialized platforms like Biohack.me are bustling marketplaces of ideas. Here, a biohacker in Tokyo can post their latest blood panel results at three in the morning and receive feedback from colleagues in Berlin and California within hours. These networks enable the real-time crowdsourcing of science, where complex problems are deconstructed and analyzed by a global brain trust with a dizzying array of expertise. Members share detailed protocols, raw data, code for analyzing genetic information, and reviews of everything from supplement brands to gene testing services.

This digital ecosystem is unparalleled for its speed and scale of information dissemination. A breakthrough in a garage lab in Sweden can be replicated and validated in community labs across the world within weeks, a pace unimaginable in traditional academic publishing. However, this strength is also a vulnerability. The lack of formal gatekeeping means misinformation can spread just as rapidly as valid insight. A charismatic individual promoting an unproven and potentially dangerous intervention can garner a significant following without the checks of peer review. Therefore, these communities rely heavily on a culture of citation, debate, and reputation. Seasoned members often act as moderators, challenging claims and demanding evidence, thus creating an organic system of quality control. The digital network ensures that no biohacker is truly alone, providing a continuous stream of peer support, technical advice, and the shared excitement of discovery that fuels a movement built on curiosity and a profound desire for self-determination.

8.3 OPEN-SOURCE BIOLOGY AND CITIZEN SCIENCE

8.3.1 Foundations of Open-Source Biology

Open-source biology is rooted in the principles of transparency, collaboration, and accessibility. Much like the open-source software movement, which revolutionized computing by making code freely available for modification and redistribution, open-source biology seeks to make biological research tools, protocols, and data accessible to the public. This approach accelerates innovation by allowing researchers, hobbyists, and citizen scientists to build upon each other's work without the constraints of proprietary restrictions.

The rise of affordable genetic engineering tools, such as CRISPR kits and DIY bioreactors, has been instrumental in this shift. Companies and organizations now offer low-cost, user-friendly biological experimentation kits, enabling individuals to conduct experiments that were once confined to professional laboratories. Additionally, open-access databases and online repositories provide free genetic sequences, research papers, and experimental protocols, further lowering the barriers to entry for aspiring biohackers.

8.3.2 Citizen Science and the Democratization of Research

Citizen science plays a crucial role in the biohacking movement by engaging non-professional scientists in meaningful research. This participatory approach allows individuals to contribute to scientific discovery, whether through data collection, experimentation, or collaborative problem-solving. Projects such as the Foldit protein-folding game, where players help solve complex biological puzzles, demonstrate how public involvement can lead to breakthroughs that might otherwise be unattainable through traditional research methods alone.

The democratization of science through citizen participation has several key benefits. It increases public engagement with scientific concepts, fosters a sense of ownership over research, and accelerates the pace of discovery by distributing the workload across a broader population. Moreover, it challenges the notion that scientific progress is solely the domain of institutional researchers, proving that innovation can emerge from diverse and decentralized sources.

8.3.3 Structure and Ethos of Biohacking Communities

Biohacking communities are built on principles of collaboration, experimentation, and mutual support. These groups often organize around shared physical spaces, such as community labs or makerspaces, where members can access equipment, share knowledge, and work on projects together. Unlike traditional research environments, which may be hierarchical and competitive, biohacking communities tend to be egalitarian, with an emphasis on peer-to-peer learning and open exchange of ideas.

One of the defining characteristics of these communities is their emphasis on DIY biology, where individuals take a hands-on approach to experimentation. This ethos encourages creativity and problem-solving, as biohackers often work with limited resources, repurposing everyday materials for scientific use. The culture also values transparency and reproducibility, with many biohackers documenting their experiments in detail and sharing their findings openly, whether through blogs, social media, or academic preprint servers.

8.3.4 Community Labs and Makerspaces

Community labs, often referred to as DIYbio labs or hackerspaces, serve as the physical hubs for biohacking activities. These spaces provide access to essential equipment such as PCR machines, centrifuges, and microscopes, which would otherwise be prohibitively expensive for individuals to purchase. They also offer workshops, training sessions, and collaborative projects that allow members to develop their skills and contribute to collective research efforts. Some well-known community labs include Genspace in New York, BioCurious in California, and La Paillasse in Paris. These organizations not only facilitate scientific experimentation but also foster a sense of community among biohackers, providing a supportive environment where ideas can be freely exchanged and refined. The existence of these labs is a testament to the growing demand for accessible, hands-on scientific engagement outside traditional academic settings.

8.3.5 Online Platforms and Knowledge Sharing

In addition to physical spaces, online platforms play a crucial role in connecting biohackers and facilitating knowledge exchange. Websites such as OpenWetWare, DIYbio.org, and various subreddits dedicated to biohacking serve as digital gathering places where individuals can share protocols, discuss experiments, and seek advice from more experienced practitioners. These platforms often host open-source repositories of genetic sequences, experimental designs, and software tools that can be freely used and modified. Social media has also become an important tool for biohackers to document their work and engage with a global audience. Platforms like Twitter, YouTube, and TikTok allow biohackers to share their progress in real time, demystifying complex biological concepts and making science more relatable to the general public. This digital presence helps to normalize biohacking as a legitimate and valuable form of scientific inquiry.

8.3.6 Ethical Considerations and Regulatory Challenges

While the biohacking movement is driven by a spirit of innovation and accessibility, it also raises important ethical and regulatory questions. The ability to conduct genetic experiments outside of controlled laboratory environments introduces potential risks, such as the accidental release of modified organisms or the misuse of biotechnology for harmful purposes. These concerns have led to debates about how biohacking should be regulated, if at all, and what safeguards should be put in place to prevent misuse.

Some biohacking communities have taken proactive steps to address these issues by establishing their own ethical guidelines and safety protocols. For example, many community labs require members to undergo training in biosafety and ethical conduct before they can use certain equipment or participate in specific projects. Additionally, organizations like the Biohacker Code of Ethics provide frameworks for responsible experimentation, emphasizing principles such as transparency, non-maleficence, and respect for life.

8.3.7 Cultural Impact of Biohacking

The biohacking movement is more than just a scientific endeavor; it represents a cultural shift in how society views science, technology, and personal agency. By empowering individuals to take an active role in biological research, biohacking challenges the traditional boundaries between experts and the public. This democratization of science fosters a culture of curiosity, experimentation, and self-improvement, where individuals are encouraged to explore and innovate without waiting for institutional approval.

This cultural shift is also reflected in the broader maker movement, which celebrates creativity, problem-solving, and hands-on engagement with technology. Biohacking aligns with this ethos by promoting the idea that anyone can contribute to scientific progress, regardless of their formal training or background. As a result, the movement has attracted a diverse range of participants, from artists and engineers to healthcare professionals and students, all united by a shared passion for exploration and discovery.

8.4 PROFILES OF NOTABLE BIOHACKERS AND INNOVATORS

8.4.1 Josiah Zayner: The Rebel Scientist

Josiah Zayner has become one of the most recognizable figures in biohacking due to his provocative experiments and public demonstrations of genetic engineering. With a background in molecular biology and professional experience at NASA, Zayner transitioned from institutional science to independent biohacking, seeking to democratize access to genetic tools.

His company, The Odin, provides do-it-yourself biotechnology kits that allow enthusiasts to explore genetic modification at home. Zayner is most widely known for experimenting on himself, including attempts to alter his own genome using CRISPR technology. His work raises both excitement and controversy, as it highlights the potential for individuals to conduct advanced genetic experiments outside of laboratories but also prompts ethical and regulatory debates about the risks of self-experimentation. Zayner represents the rebellious spirit of biohacking, combining entrepreneurial drive with a challenge to conventional boundaries of science.

8.4.2 Ellen Jorgensen: The Advocate for Community Science

Ellen Jorgensen stands out as one of the most respected advocates for community-driven biology. As the co-founder of Genspace in New York City, one of the first community biology laboratories in the world, she has dedicated her career to making science accessible to the public. With a PhD in molecular biology and years of research experience, Jorgensen recognized the need to create spaces where non-scientists could engage with biotechnology in safe and supportive environments. Under her leadership, Genspace has become a model for community labs, offering workshops, collaborative projects, and public education initiatives. Jorgensen's work emphasizes transparency, biosafety, and inclusivity, setting a standard for ethical engagement in biohacking. Her contributions illustrate how biohacking can move beyond self-experimentation to serve as a platform for education, creativity, and civic science.

8.4.3 Dave Asprey: The Father of Biohacking Culture

Dave Asprey approaches biohacking from the perspective of health optimization and personal performance. Known widely as the founder of Bulletproof, he has popularized the idea of biohacking as a lifestyle centered on nutrition, supplementation, and technological interventions. Asprey's philosophy blends scientific research with entrepreneurial branding, offering products and techniques designed to enhance energy, focus, and longevity.

His promotion of Bulletproof Coffee and emphasis on mitochondrial health have influenced millions of followers who identify as biohackers in pursuit of improved well-being. Although some critics argue that his claims lack rigorous scientific validation, Asprey's role in spreading biohacking culture is undeniable. By framing biohacking as accessible to anyone seeking self-improvement, he transformed the term from a niche scientific practice into a mainstream wellness movement.

8.4.4 Amal Graafstra: Pioneer of Human Augmentation

Amal Graafstra has carved a distinct place in the biohacking community as a pioneer of human augmentation through implantable technology. He is best known for developing RFID and NFC implants that allow individuals to unlock doors, store information, and interact with digital systems using embedded chips. His company, Dangerous Things, provides implant technology to enthusiasts worldwide, advancing the concept of cyborg-like integration between humans and machines. Graafstra's vision extends beyond convenience, as he advocates for a future where human bodies can be enhanced through technological interfaces. While his work has sparked debates about privacy, security, and ethics, it demonstrates the expanding horizons of biohacking into the realm of human-machine convergence. Graafstra embodies the technological side of biohacking, where biology and digital systems merge to redefine human capabilities.

8.4.5 Thomas Goetz: Bridging Data and Health

Thomas Goetz has contributed to biohacking through his efforts to democratize access to genetic information and health data. As the co-founder of the genetic testing company 23andMe and later ventures in digital health, Goetz emphasizes the importance of individuals having control over their own biological information. His work reflects the ethos of empowering people to make informed decisions about health and wellness through personal data. Goetz's career demonstrates how biohacking extends beyond experimental biology to include data-driven self-knowledge. By promoting direct-to-consumer genetic testing and personalized health platforms, he has helped shape the cultural shift toward viewing biology as something individuals can actively interpret and optimize.

8.4.6 Brian Hanley: Bold Self-Experimentation

Brian Hanley is a biohacker who gained attention for injecting himself with experimental gene therapy intended to boost immune function and longevity. A microbiologist by training, Hanley bypassed traditional clinical testing pathways in pursuit of personal enhancement, embodying the spirit of self-directed experimentation. His actions highlight both the ambition and the risks inherent in biohacking, as self-experimentation raises questions about safety, consent, and scientific validity. Hanley's work is controversial, but it represents the drive of some biohackers to push scientific frontiers through personal risk-taking. His example underscores the tension between innovation and ethics in the broader movement, illustrating how biohacking often challenges regulatory norms.

8.4.7 Gabriel Licina: The Visionary of Human Enhancement

Gabriel Licina, co-founder of the science collective Science for the Masses, gained international attention for experimenting with chlorophyll-like compounds intended to enhance human night vision. Licina represents the exploratory and unconventional side of biohacking, where the goal is not only to optimize health but also to expand human abilities beyond natural limits. His willingness to test bold hypotheses reflects the experimental ethos of grassroots science, where breakthroughs may come from outside traditional research structures. Although his night vision experiment generated skepticism, it also demonstrated the imagination and audacity that drive biohackers to explore possibilities that conventional science might dismiss. Licina embodies the boundary-pushing nature of biohacking as a field that blends curiosity, risk, and vision.

8.4.8 Zoltan Istvan: Transhumanist Advocate

Zoltan Istvan, a journalist, author, and political figure, is closely associated with the transhumanist movement, which overlaps with biohacking in its focus on human enhancement. Through his writings and activism, Istvan has championed the use of biotechnology, artificial intelligence, and nanotechnology to extend human life and capabilities. His presidential campaign in the United States under the banner of the Transhumanist Party brought public attention to ideas once considered fringe, including radical life extension and human-machine integration. While not a laboratory biohacker, Istvan's influence lies in framing biohacking within a broader cultural and philosophical context. He emphasizes the societal implications of human enhancement and advocates for policies that embrace technological progress as a means of improving human existence.

8.5 ROLE OF SOCIAL MEDIA AND CONTENT PLATFORMS

8.5.1 Digital Ecosystem of Biohacking Communities

Biohacking communities have flourished in the digital age, leveraging social media and content platforms to create interconnected networks of enthusiasts, researchers, and citizen scientists. These platforms serve multiple functions: they act as educational hubs, discussion forums, and spaces for showcasing experimental results. Unlike traditional scientific publishing, which can be slow and exclusive, social media allows for instant sharing of ideas, troubleshooting, and peer feedback.

Platforms like YouTube have become essential for visual learning, where biohackers document their experiments, explain complex biological concepts, and demonstrate DIY techniques. Twitter and Instagram facilitate quick updates, discussions, and networking among biohackers, while Reddit and Discord serve as more in-depth discussion forums where detailed technical questions can be explored. Additionally, specialized platforms like OpenWetWare and DIYbio.org provide structured repositories for protocols, research papers, and collaborative projects.

8.5.2 Social Media as an Educational Tool

One of the most significant contributions of social media to biohacking culture is its role in education. Traditional biological research often requires years of formal training, but social media has made it possible for individuals to learn through self-directed study. YouTube channels dedicated to biohacking, such as those run by The Thought Emporium or Amino Labs, break down complex topics like CRISPR gene editing, synthetic biology, and DIY lab techniques into digestible, engaging content. Similarly, TikTok and Instagram Reels have introduced shorter, more accessible explanations of biological concepts, making them appealing to younger audiences. These platforms help demystify science, showing that biohacking is not just for professionals but for anyone with curiosity and determination. The visual and interactive nature of social media makes learning more engaging, allowing viewers to see experiments in action rather than just reading about them in textbooks.

8.5.3 Fostering Collaboration and Crowdsourced Science

Social media and content platforms have also enabled unprecedented levels of collaboration within the biohacking community. Unlike traditional research, which often operates in silos, biohackers use digital spaces to crowdsource solutions, share data, and refine experimental techniques. For example, a biohacker struggling with a PCR protocol might post their issue on a Reddit forum and receive troubleshooting advice from experienced practitioners within hours.

Platforms like GitHub, traditionally used for software development, have also been adopted by biohackers to share genetic sequences, lab protocols, and open-source hardware designs. This collaborative approach accelerates innovation by allowing individuals to build upon each other's work, much like the open-source software movement. The result is a more dynamic and iterative scientific process, where ideas evolve rapidly through collective input.

8.5.4 Influence of Content Creators and Thought Leaders

The biohacking movement has been shaped by influential content creators who use social media to inspire and educate. Figures like Josiah Zayner, a prominent biohacker and science communicator, have used platforms like YouTube and Twitter to document their experiments, advocate for open science, and challenge regulatory barriers. Their work has helped normalize biohacking as a legitimate form of scientific inquiry, encouraging others to explore DIY biology.

These thought leaders often serve as bridges between the biohacking community and the broader public, translating complex scientific concepts into accessible language. Their content not only educates but also fosters a sense of community, making biohacking feel like an inclusive and exciting endeavor rather than an exclusive academic pursuit. By humanizing science, they help break down the intimidation factor that often surrounds biological research.

CHAPTER 9

Risks, Ethics, and Regulations

9.1 HEALTH RISKS AND ADVERSE EFFECTS

9.1.1 Dangers of Genetic and DIY Biology Experiments

One of the most controversial and high-risk areas of biohacking involves genetic experimentation, particularly with tools like CRISPR-Cas9. While CRISPR has revolutionized genetic research, its use outside controlled laboratory settings poses significant dangers. DIY genetic modification, such as self-administered gene therapy or attempts to edit one's own DNA, carries the risk of unintended mutations, off-target effects, and potential long-term health consequences. Unlike professional researchers who work in sterile environments with strict safety protocols, amateur biohackers may lack the necessary expertise to handle genetic material safely, increasing the likelihood of contamination or accidental exposure to harmful substances.

Furthermore, the ethical implications of genetic biohacking are profound. Modifying one's own genes—or those of others—without fully understanding the consequences could lead to irreversible changes, including unintended hereditary effects. The potential for misuse, such as creating harmful pathogens or altering germlines, also raises serious biosecurity concerns. While some biohackers argue that self-experimentation accelerates scientific progress, the lack of regulatory oversight means that the risks often outweigh the benefits, particularly when dealing with complex biological systems.

9.1.2 Unregulated Supplement and Nootropic Use

Another common biohacking practice involves the use of supplements, nootropics, and performance-enhancing compounds, many of which are not approved by regulatory agencies like the FDA. While some of these substances may offer cognitive or physical benefits, their long-term effects remain poorly understood. The lack of standardized dosing, potential interactions with medications, and the risk of contamination in unregulated products pose significant health hazards. For example, nootropics like racetams or modafinil, often used to enhance focus and productivity, can lead to side effects such as insomnia, anxiety, or cardiovascular strain when misused. Similarly, experimental peptides and hormones, sometimes sourced from questionable suppliers, may cause hormonal imbalances, organ damage, or other severe health complications. The biohacking community's emphasis on self-optimization can sometimes lead to reckless experimentation, where individuals prioritize short-term gains over long-term safety.

9.1.3 Implantable Technologies and Bodily Modifications

The growing trend of implantable technologies, such as RFID chips, magnetic implants, and neural interfaces, introduces another layer of risk. While these devices can offer convenience—such as unlocking doors with a hand wave or monitoring biometric data—they also come with potential health complications. Surgical implantation carries risks of infection, nerve damage, and immune rejection, particularly when performed outside medical settings.

Moreover, the long-term effects of having foreign materials embedded in the body are not fully understood. Some implants may degrade over time, leading to inflammation or toxicity. Others could interfere with medical imaging or cause unforeseen interactions with the body's natural processes. The lack of standardized safety protocols in DIY biohacking means that individuals who undergo these procedures often do so without proper medical supervision, increasing the likelihood of adverse outcomes.

9.1.4 Extreme Lifestyle and Dietary Biohacking

Biohacking often extends to extreme lifestyle modifications, including radical diets, fasting protocols, and sleep optimization techniques. While some of these practices, such as intermittent fasting or ketogenic diets, have scientific backing, others—like prolonged water fasting or extreme caloric restriction—can lead to malnutrition, metabolic disorders, and other health complications. The biohacking community's tendency to push biological limits can sometimes result in dangerous experimentation, where individuals disregard established medical guidelines in pursuit of optimization.

Additionally, the use of biofeedback devices and wearable technology, while beneficial for tracking health metrics, can also lead to obsessive self-monitoring. This hyper-focus on data-driven optimization may contribute to anxiety, stress, and an unhealthy relationship with one's body. The pressure to constantly "upgrade" oneself can create psychological strain, where individuals feel compelled to pursue ever-more-extreme modifications to achieve perceived perfection.

9.1.5 Ethical and Legal Implications of Biohacking Risks

The health risks associated with biohacking extend beyond individual consequences to broader ethical and legal concerns. The lack of regulation in the biohacking space means that many practices operate in a legal gray area, where individuals may unknowingly violate biosafety laws or medical ethics. For instance, self-administered gene therapy or the distribution of untested biological materials could have legal repercussions, particularly if they result in harm to oneself or others. Ethically, the biohacking movement raises questions about informed consent, particularly when individuals experiment on themselves without fully understanding the risks. The potential for exploitation also exists, where companies or influencers may promote unproven or dangerous biohacks for profit, putting vulnerable individuals at risk. The decentralized nature of biohacking makes it difficult to enforce safety standards, leaving participants to navigate these risks largely on their own.

9.1.6 Need for Education and Responsible Innovation

Given the inherent risks of biohacking, there is a pressing need for education and responsible innovation within the community. Many adverse effects stem from a lack of proper training, misinformation, or the pursuit of quick results without considering long-term consequences. Establishing clear safety guidelines, promoting evidence-based practices, and encouraging peer-reviewed experimentation could help mitigate some of these dangers.

Community labs and biohacking organizations have a role to play in fostering a culture of responsibility. By providing access to proper equipment, mentorship, and ethical frameworks, these spaces can help ensure that experimentation is conducted safely and transparently. Additionally, greater collaboration with medical professionals and regulatory bodies could help bridge the gap between DIY biohacking and established scientific research, reducing the likelihood of harmful outcomes.

9.2 LEGAL BOUNDARIES OF SELF-EXPERIMENTATION

9.2.1 Medical Practice Act: Practicing Medicine on Oneself

One of the most significant legal gray areas involves the application of Medical Practice Acts. These state laws prohibit the unlicensed practice of medicine, which is typically defined as diagnosing, treating, curing, or preventing disease. A biohacker who self-administers an experimental gene therapy intended to mitigate a genetic condition or who uses a DIY fecal microbiota transplant to address a digestive disorder is, in a literal sense, engaging in an act of treatment. The critical legal question becomes whether treating oneself constitutes the "practice of medicine" as defined by law. Most statutes were written to protect the public from unqualified practitioners, not to regulate an individual's actions upon their own body. Consequently, there is precedent for prosecuting an individual solely for self-treatment. The legal risk escalates dramatically, however, the moment a biohacker provides advice or assistance to another person. Sharing a protocol, administering a treatment to someone else, or even providing the materials for another's self-experimentation could easily be construed as practicing medicine without a license, opening the individual to serious criminal and civil liability. This creates a paradox where an action may be legally ambiguous when performed on oneself but becomes clearly illegal when shared within the community, the very lifeblood of the biohacking ethos.

9.2.2 FDA and the Regulation of Biologics and Devices

The jurisdiction of the U.S. Food and Drug Administration (FDA) introduces another layer of complexity. The FDA regulates drugs, biologics (including gene therapies), and medical devices to ensure they are safe and effective for their intended use. The agency's authority is triggered when a product is introduced into interstate commerce. A biohacker who manufactures a substance—for example, by compounding a novel nootropic blend or creating a gene therapy vector in a home lab—for their own personal use occupies a contested space. The FDA has historically taken the position that the Federal Food, Drug, and Cosmetic Act does not explicitly exempt self-directed experimentation. However, enforcement against individuals for personal use is an extremely low priority, focusing instead on entities selling unapproved products to the public. The legal line is crossed when there is any form of distribution. Selling, trading, or even giving away an unapproved drug, biologic, or device is a direct violation of FDA regulations. This poses a significant constraint on the biohacking community's collaborative spirit. A biohacker cannot legally mail a sample of their DIY biologic to a fellow enthusiast for independent verification without potentially violating federal law, stifling the peer review and replication that are cornerstones of scientific progress.

9.2.3 Liability and the Consequences of Failure

Beyond direct regulatory action, biohackers face significant potential liability under civil law, particularly the law of negligence. If a self-experiment goes awry and the individual requires emergency medical care, the resulting costs could be catastrophic. More importantly, if a biohacker involves another person, even tangentially, they open themselves to lawsuits. If an individual follows a protocol shared online and suffers injury, the author of that protocol could potentially be sued for damages under a theory of negligence.

The plaintiff would need to establish that the biohacker owed a duty of care, breached that duty by providing unreasonably dangerous or inaccurate information, and that this breach directly caused the injury. While winning such a case might be challenging, the financial and emotional cost of defending against it could be devastating. Furthermore, if a self-experiment results in harm to a third party—

for example, if an attempt at cognitive enhancement leads to impaired judgment that causes a car accident—standard personal liability principles would apply. The biohacker's intentional deviation from the norm could be used as evidence of recklessness, potentially negating insurance coverage and exposing personal assets to seizure.

9.2.4 Intellectual Property and Bio-Security Concerns

The legal landscape extends beyond direct health regulation into the realms of intellectual property and national security. Many of the tools and techniques fundamental to advanced biohacking, such as CRISPR-Cas9 systems, are protected by powerful patents. While patent holders rarely pursue individual tinkerers, a community lab that uses patented biological components without a license could become a target for litigation, threatening the viability of these essential spaces.

Even more serious are the regulations enforced by the Federal Select Agent Program, which governs the possession, use, and transfer of biological agents and toxins deemed to pose a severe threat to public health. While most biohackers steer clear of these dangerous materials, the very act of genetic engineering can sometimes fall under heightened scrutiny. The act of synthesizing DNA is regulated, and companies that provide synthetic gene fragments are required to screen orders against sequences of concern, such as those for pathogenic viruses. A biohacker attempting to order materials for an experiment could find their order denied or, in a worst-case scenario, trigger an investigation by law enforcement agencies concerned with bioterrorism, even if their intentions are purely benign and personal.

9.3 BIOHACKING AND BIOMEDICAL ETHICS

9.3.1 Autonomy and the Right to Self-Experimentation

One of the strongest arguments in favor of biohacking is the principle of autonomy. Advocates argue that individuals should have the right to control their bodies and experiment with interventions that may improve their health, capabilities, or lifespan. Self-experimentation has a long history in science, with many breakthroughs achieved through personal risk-taking. Biohackers extend this tradition by applying it in decentralized, non-institutional contexts. However, autonomy must be balanced with considerations of safety and informed consent. When experiments are performed without adequate understanding of risks, the ethical justification weakens. The challenge for biomedical ethics is determining how far the right to self-experimentation extends, and whether personal autonomy can justify interventions that may carry serious consequences not only for the individual but also for public health.

9.3.2 Safety and the Risk of Harm

Safety is perhaps the most urgent ethical concern associated with biohacking. Traditional biomedical research operates under strict protocols designed to minimize harm and ensure participant protection. In contrast, many biohacking practices take place outside regulated frameworks, leaving individuals vulnerable to unforeseen consequences.

For example, genetic self-experimentation carries risks of unintended mutations, infections, or immune reactions, while implantable devices may cause complications ranging from tissue damage to cybersecurity threats. Nutritional and pharmacological biohacking can also expose individuals to long-term health problems if interventions are poorly studied or misapplied. Biomedical ethics insists on the principle of non-maleficence, or "do no harm," yet in the realm of biohacking, this principle is difficult to enforce. The question remains whether it is ethical to allow unrestricted experimentation when the risks extend beyond the individual, potentially endangering others through the spread of biological materials or misuse of technologies.

9.3.3 Justice and Accessibility

The principle of justice in biomedical ethics demands fairness in the distribution of medical resources and access to health benefits. Biohacking introduces new inequalities, as access to advanced technologies, implants, or genetic testing often depends on wealth and location. While community laboratories attempt to democratize science, cutting-edge interventions such as personalized genetic editing remain accessible only to the privileged.

This creates ethical concerns about a future where human enhancement deepens social divides, producing a class of biologically advantaged individuals. At the same time, biohacking could also serve as a force for equity if open-source communities succeed in making knowledge and tools widely available. Biomedical ethics must grapple with how to ensure that the benefits of biohacking are not concentrated in the hands of a few but distributed in ways that promote fairness and inclusivity.

9.3.4 Responsibility and Regulation

Responsibility is a central issue when considering biohacking. In traditional medicine, physicians, researchers, and institutions bear responsibility for patient safety and research outcomes. In biohacking, responsibility is far more diffuse. Who should be held accountable when a self-administered genetic experiment goes wrong? Should governments regulate personal biohacking as they do pharmaceuticals, or would this infringe upon individual freedoms? Overregulation risks stifling innovation, while underregulation leaves society vulnerable to harm. Some biohacking communities have attempted to self-regulate by creating codes of conduct and biosafety standards, but these measures lack the enforcement power of formal oversight. Biomedical ethics highlights the need for clear frameworks of responsibility that respect autonomy while safeguarding public health and scientific integrity.

9.3.5 Ethical Debate over Human Enhancement

Beyond immediate health and safety concerns, biohacking raises deeper philosophical questions about the ethics of human enhancement. Should individuals be allowed to alter their biology to improve memory, strength, or longevity? Biomedical ethics has traditionally grappled with the line between therapy and enhancement, with therapy seen as restoring normal function and enhancement as moving beyond it. Biohacking blurs this distinction by treating enhancement as a personal choice rather than a societal judgment. Critics argue that enhancement undermines the natural human condition and risks creating new forms of inequality. Supporters counter that restricting enhancement violates personal freedom and slows progress. The ethical debate here reflects broader questions about the future of humanity, transhumanism, and the very definition of what it means to be human.

9.3.6 Privacy and Data Ethics

As biohacking increasingly intersects with digital health technologies, issues of privacy and data ethics become paramount. Devices that monitor sleep, heart rate, or genetic information generate vast amounts of personal data. When stored on corporate servers or shared through apps, this data becomes vulnerable to misuse, surveillance, or exploitation for profit. Biomedical ethics emphasizes the principle of confidentiality, but social media and content platforms often encourage sharing of personal health data in ways that compromise privacy. For biohackers, the allure of community validation or open-source collaboration must be balanced against the risks of data exposure. Ethical frameworks must address how to protect individual privacy while enabling the collaborative spirit that drives much of the biohacking movement.

9.3.7 Social and Cultural Implications

The ethical dimensions of biohacking extend into broader social and cultural contexts. Biohacking challenges traditional medical authority, raising questions about who should have the right to define, practice, and advance biology. It also confronts cultural attitudes toward the body, naturalness, and human identity. In some societies, body modification through implants may be celebrated as innovation, while in others it may be stigmatized as unnatural. The movement's emphasis on self-directed experimentation also intersects with cultural values around independence, responsibility, and collective well-being. Biomedical ethics must take into account these diverse cultural perspectives, recognizing that the ethical evaluation of biohacking cannot be separated from the societies in which it is practiced.

9.4 MEDICAL VS. NON-MEDICAL ENHANCEMENTS

9.4.1 Defining Medical Enhancements in Biohacking

Medical enhancements in biohacking refer to interventions that are either prescribed by healthcare professionals or developed within regulated medical research environments. These enhancements are typically designed to address specific health concerns, such as hormone replacement therapy for deficiencies, neurostimulation for depression, or genetic therapies for hereditary conditions. Unlike purely experimental biohacks, medical enhancements undergo rigorous testing to ensure safety and efficacy before being approved for use.

One of the key advantages of medical enhancements is their foundation in evidence-based science. Treatments like insulin pumps for diabetics or cochlear implants for hearing loss have been extensively studied and refined, minimizing risks while maximizing benefits. Even emerging medical biohacks, such as closed-loop brain stimulation for neurological disorders, are developed under strict regulatory oversight to prevent harm. This structured approach provides a level of reliability and accountability that is often lacking in non-medical biohacking.

However, medical enhancements are not without limitations. Access to these treatments is frequently restricted by cost, regulatory approval processes, and medical necessity criteria, which can exclude individuals seeking enhancements for non-therapeutic purposes. Additionally, the conservative nature of medical research means that some potentially beneficial enhancements may take years to reach the public, if they are approved at all.

9.4.2 Rise of Non-Medical Enhancements

Non-medical enhancements represent a more experimental and often controversial side of biohacking. These modifications are pursued not for health reasons but for performance optimization, aesthetic improvements, or personal curiosity. Examples include nootropic use for cognitive enhancement, magnetic implants for sensory augmentation, and CRISPR-based gene editing for non-therapeutic traits. Unlike medical enhancements, these practices often operate in regulatory gray areas, where safety and long-term effects are not well understood.

The appeal of non-medical enhancements lies in their potential to push human capabilities beyond natural limits. Biohackers who experiment with smart drugs, biofeedback devices, or even DIY genetic modifications are often motivated by a desire for self-improvement rather than medical necessity. This approach aligns with the broader transhumanist philosophy, which advocates for using technology to transcend biological constraints. However, the lack of clinical oversight means that these enhancements carry significant risks, including unintended side effects, long-term health consequences, and ethical concerns about fairness and accessibility.

9.4.3 Safety and Efficacy: A Comparative Analysis

When comparing medical and non-medical enhancements, safety and efficacy emerge as critical differentiating factors. Medical enhancements are subject to clinical trials, regulatory approvals, and post-market monitoring, which help ensure that risks are minimized and benefits are maximized. In contrast, non-medical enhancements often lack this level of scrutiny, relying instead on anecdotal evidence, self-reporting, and community-driven experimentation.

For instance, a medically approved nootropic like modafinil, prescribed for narcolepsy, has undergone extensive testing to confirm its safety profile and therapeutic effects. Meanwhile, a non-medical nootropic stack purchased online may contain unregulated substances with unknown interactions or contaminants. Similarly, a surgically implanted medical device, such as a pacemaker, follows strict safety protocols, whereas a DIY magnetic implant carries risks of infection, rejection, or unintended physiological effects.

This disparity in safety standards highlights a fundamental tension in biohacking: the pursuit of enhancement must be balanced with responsible risk assessment. While medical enhancements provide a more secure pathway to improvement, non-medical enhancements offer greater freedom and experimentation at the cost of higher uncertainty.

9.4.4 Ethical Considerations and Societal Impact

The ethical implications of medical versus non-medical enhancements are complex and multifaceted. Medical enhancements are generally viewed as ethically justifiable when they restore or improve health, but debates arise when they are used for non-therapeutic purposes, such as cosmetic gene editing or cognitive enhancement in healthy individuals. Non-medical enhancements, on the other hand, often face greater ethical scrutiny due to their experimental nature and potential for misuse.

One major ethical concern is the issue of fairness and accessibility. Medical enhancements, while regulated, are often expensive and available only to those with financial means or insurance coverage. Non-medical enhancements, while sometimes more affordable, can create disparities where only certain individuals have access to performance-boosting technologies, leading to societal inequalities. Additionally, the pursuit of non-medical enhancements raises questions about bodily autonomy, consent, and the potential for coercion in competitive environments like workplaces or sports.

Another ethical dilemma involves the long-term consequences of enhancement technologies. Medical enhancements are typically designed with patient safety in mind, but non-medical modifications may have unforeseen effects on future generations, particularly in the case of heritable genetic edits. The lack of long-term data on many non-medical enhancements means that individuals may be taking risks not only for themselves but also for their descendants.

9.5 INSTITUTIONAL AND GOVERNMENTAL PERSPECTIVES

9.5.1 Academic Research Establishment: From Skepticism to Cautious Collaboration

The traditional academic world views biohacking through a lens of deep-seated skepticism, and often, outright disdain. The core values of academia—peer review, rigorous methodology, controlled environments, and institutional oversight—stand in direct contrast to the biohacker's embrace of trial-and-error, self-experimentation, and decentralized discovery. To many established scientists, biohacking projects lack the necessary controls and statistical power to generate valid, reproducible knowledge. The practice of self-experimentation is particularly anathema, violating the ethical cornerstone of informed consent as governed by Institutional Review Boards (IRBs). From an academic perspective, the biohacker is an unqualified amateur whose work risks generating misinformation, causing personal harm, and eroding public trust in science. This skepticism is compounded by a professional culture that prioritizes publication in high-impact journals, a system that inherently devalues work conducted outside its walls. However, a more nuanced perspective is emerging within some academic circles. Forward-thinking researchers and institutions recognize that the biohacking community represents a vast, untapped reservoir of creativity, passion, and diverse skill sets. Initiatives are slowly developing to bridge this divide. Some universities are opening their doors to community lab collaborations, providing access to high-end equipment and expert mentorship in exchange for fresh ideas. Others are studying the biohacking movement itself as a fascinating sociological phenomenon, a citizen-science rebellion against the ivory tower. This shift from pure skepticism to cautious collaboration acknowledges that innovation can indeed spring from unexpected sources, even if it requires translation to fit within the academic paradigm.

9.5.2 Pharmaceutical and Biotechnology Industry: Disruption and Defensive Innovation

The perspective of the multi-trillion-dollar pharmaceutical and biotech industry is primarily one of perceived threat and potential disruption. The industry's business model is built on a foundation of proprietary research, extensive clinical trials, and patent protection, a process that takes over a decade and costs billions of dollars per approved therapy. Biohacking, with its open-source ethos and direct-to-consumer approach, challenges every aspect of this model. The industry fears the rise of "garage biotech" that could develop and distribute therapies outside its control, potentially endangering consumers and undermining the economic rationale for their massive R&D investments. This perspective fuels strong opposition to many forms of biohacking, particularly those involving unapproved genetic therapies or drug compounding. Industry lobbyists actively advocate for stricter regulations on genetic sequencing services, synthetic DNA providers, and lab equipment sales to private individuals, arguing that such measures are necessary for public safety.

Yet, simultaneously, a more opportunistic perspective exists within corporate R&D departments. Just as tech giants acquire promising startups, large biotech firms are keenly aware of the innovative potential within the biohacking community. They engage in defensive innovation, monitoring biohacking forums and publications for breakthrough ideas that can be patented, developed, and commercialized through traditional channels. Some companies have even launched "innovation challenges" or grant programs aimed at biohackers, effectively seeking to co-opt the movement's creativity and bring it back within the proprietary fold. The industry's ultimate goal is to harness the energy of biohacking without surrendering control over the product pipeline or the regulatory landscape.

9.5.3 Governmental Regulatory Agencies: The Impossible Mandate

Government regulatory agencies, most notably the U.S. Food and Drug Administration (FDA), face an almost impossible mandate in relation to biohacking. Their primary mission is to protect public health by ensuring the safety and efficacy of drugs, biologics, and medical devices. The biohacking movement, by its very nature, operates in spaces these regulations were not designed to cover. The FDA's existing framework is ill-equipped to handle an individual who manufactures and self-administers a therapy in their own home. The agency's perspective is therefore necessarily cautious and reactive. It focuses its enforcement resources on clear public threats, such as companies or individuals selling unapproved, adulterated, or misbranded medical products directly to consumers.

High-profile cases, such as the warning letters sent to companies selling DIY genetic therapy kits, serve as public demarcations of the agency's boundaries. However, the agency remains largely silent on pure self-experimentation, creating a de facto but precarious tolerance. This ambiguity is fraught with difficulty. Too heavy-handed an approach could be seen as infringing on personal liberty and stifling innovation; too lenient an approach could be condemned as negligent if a public health crisis emerges from a biohacking incident. The agency is slowly evolving its perspective, exploring new regulatory models like the "right to try" legislation and examining how to adapt its oversight to accommodate rapid innovation while maintaining its core protective function. This process is slow, deliberate, and fraught with legal and ethical challenges, reflecting the government's struggle to apply twentieth-century laws to twenty-first-century science.

9.5.4 Law Enforcement and National Security: The Biosecurity Dilemma

For law enforcement and national security agencies, the perspective on biohacking is dominated by the specter of biosecurity risk. Agencies like the FBI and the Department of Homeland Security view the democratization of biotechnology through a security lens, where the same tools and knowledge used for benevolent self-experimentation could be misappropriated by malicious actors for bioterrorism or biowarfare. The core of their concern is the "dual-use dilemma": the fact that a technique intended to help engineer yeast for sustainable fuel could also be used to engineer a pathogen. This perspective leads to a stance of heightened surveillance and preemptive intervention. The FBI has established dedicated outreach programs, such as the Biological Countermeasures Unit, to engage with the DIY bio community. These efforts aim to build relationships, promote a culture of responsibility, and identify potential bad actors before they act. While framed as collaboration, the underlying purpose is one of monitoring and risk mitigation. This security-focused perspective often clashes with the biohacking community's culture of transparency and open access. Security agencies may view the online sharing of a successful protocol for bypassing a gene synthesis company's screening protocol as a dangerous act, while the community sees it as a triumph of open science. This tension between security and freedom defines the governmental perspective on the most advanced and powerful forms of biohacking, ensuring that it will remain a contentious and closely watched field.

CHAPTER 10

The Future of Biohacking

10.1 EMERGING TECHNOLOGIES: AI, BCI, AND SYNTHETIC BIOLOGY

10.1.1 Artificial Intelligence in Biohacking

Artificial intelligence has become central to biohacking because of its unparalleled ability to analyze complex biological data. AI algorithms are increasingly used to process genetic information, interpret personal health records, and provide real-time feedback on physiological states. For biohackers, this means gaining insights into how their bodies function with a level of detail that was once reserved for professional laboratories and medical institutions. Personalized nutrition and fitness strategies can now be optimized with AI-driven platforms that track biomarkers such as glucose levels, heart rate variability, and sleep patterns. These tools not only enhance individual performance but also allow biohackers to design precise interventions for health and longevity.

AI also plays a vital role in predictive modeling, enabling users to anticipate potential health issues before they manifest. By analyzing genomic data and environmental factors, artificial intelligence can help identify predispositions to diseases and suggest preventive measures. This transforms biohacking from reactive self-experimentation to proactive health management. Moreover, AI-driven drug discovery platforms allow individuals and open-source communities to experiment with novel compounds or repurpose existing ones, expanding the horizon of do-it-yourself biomedical innovation. The integration of AI into biohacking represents a shift toward data-driven personalization, where human biology can be fine-tuned with unprecedented precision.

10.1.2 Brain-Computer Interfaces and Human Enhancement

Among the most futuristic technologies linked to biohacking are brain-computer interfaces, or BCIs. These systems establish direct communication pathways between the brain and external devices, bridging the gap between human cognition and digital systems. While much of the research into BCIs has been conducted in clinical contexts, such as restoring communication for paralyzed patients or assisting individuals with neurological disorders, biohackers are increasingly interested in their potential for cognitive enhancement.

BCIs promise to blur the line between biological and artificial intelligence, allowing humans to augment memory, attention, and problem-solving capabilities by interfacing with computational systems. For example, a BCI could enable a biohacker to control devices directly through thought, enhance learning speed, or expand sensory perception by creating new forms of input that the brain can interpret. Some experimental projects already explore non-invasive BCIs through wearable headsets, while more radical approaches envision implantable devices that can permanently extend human capabilities.

The appeal of BCIs within biohacking lies not only in enhancing mental performance but also in democratizing access to neurological insights. By monitoring brainwave activity, individuals can experiment with methods to improve focus, regulate emotions, or optimize sleep cycles. These applications reflect the ethos of biohacking, where individuals seek to unlock greater control over their bodies and minds. However, BCIs also raise profound ethical questions about privacy, autonomy, and the merging of human identity with machines. If brain signals can be read and interpreted by external systems, issues of consent and security become as critical as the technology itself.

10.1.3 Synthetic Biology and the Redesign of Life

Perhaps the most transformative of all emerging technologies in the biohacking domain is synthetic biology, which goes beyond modifying existing organisms to designing and constructing new forms of life. Synthetic biology applies engineering principles to biology, enabling the creation of organisms with tailored functions, from bacteria that produce biofuels to yeast that synthesizes pharmaceuticals. For biohackers, synthetic biology offers a playground of limitless possibilities, where the genetic code can be rewritten to explore new forms of biological expression.

DIY biology laboratories around the world are experimenting with genetic engineering kits that allow non-specialists to manipulate DNA and create new biological systems. Synthetic biology thus extends biohacking beyond the individual body into the broader manipulation of living systems. For instance, biohackers can experiment with engineered microbes that alter the gut microbiome for health benefits, or design organisms that produce nutrients more efficiently. This blurring of the line between laboratory science and personal experimentation illustrates how synthetic biology embodies the core spirit of biohacking: accessibility, innovation, and radical curiosity. The implications of synthetic biology are profound. It opens the door to custom biological solutions for global challenges, including food security, renewable energy, and disease treatment. At the same time, it raises significant safety concerns. Unintended consequences of synthetic organisms escaping into the environment or being misused for harmful purposes represent some of the most pressing ethical challenges in this field. While synthetic biology empowers individuals and communities, it also demands a heightened sense of responsibility and awareness of the global impact of local experiments.

10.1.4 Intersections of Emerging Technologies

What makes AI, BCIs, and synthetic biology particularly powerful in the context of biohacking is the way they intersect and reinforce each other. Artificial intelligence can accelerate discoveries in synthetic biology by analyzing genetic data and predicting the outcomes of modifications. Brain-computer interfaces can benefit from AI systems that interpret brain signals more accurately, translating thoughts into actions with higher efficiency. Synthetic biology, in turn, may support BCIs by developing biological interfaces that seamlessly integrate with neural tissue. Together, these technologies create a synergistic ecosystem of human enhancement and self-directed experimentation that pushes the boundaries of what biohacking can achieve.

This convergence also highlights the future trajectory of biohacking as not merely a collection of separate practices but as a deeply integrated system of technologies that connect biology, computation, and human agency. The possibilities range from personalized medicine that dynamically adapts to an individual's genetic profile, to hybrid human-machine systems capable of unprecedented cognitive and physical feats. The blending of these fields signals the dawn of a new era where biohacking transcends self-optimization and enters the realm of redefining humanity itself.

10.1.5 Ethical and Societal Implications

As with all powerful technologies, the integration of AI, BCIs, and synthetic biology into biohacking raises complex ethical and societal issues. Questions about safety, privacy, and fairness become increasingly urgent as individuals gain access to tools that were once tightly controlled. The potential for misuse, whether through accidental release of engineered organisms or unauthorized access to neural data, cannot be ignored. Furthermore, the uneven distribution of these technologies risks deepening social inequalities, creating divisions between those who can afford enhancements and those who cannot.

Society must also confront philosophical questions about identity and human nature. When cognitive abilities are extended through BCIs or biological traits are redesigned through synthetic biology, what remains of the natural human experience? Supporters argue that these technologies embody the next stage of human evolution, while critics caution against the loss of authenticity and the commodification of human biology. Addressing these dilemmas requires open dialogue between biohackers, ethicists, policymakers, and the broader public, ensuring that innovation is guided by principles of responsibility and inclusivity.

10.2 PERSONALIZED AND PRECISION BIOHACKING

10.2.1 Foundation of the Quantified Self: Deep Phenotyping

The first and most critical step in precision biohacking is the comprehensive collection of personal data, a process known as deep phenotyping. This goes far beyond tracking steps or sleep hours; it involves generating a multi-layered, high-resolution map of one's current biological state. The modern biohacker's toolkit for this purpose is expansive and increasingly accessible. At the genomic level, direct-to-consumer sequencing provides insights into single nucleotide polymorphisms (SNPs) that can influence metabolism, nutrient processing, neurotransmitter balance, and susceptibility to certain health conditions.

This static genetic blueprint, however, is only the starting point. Dynamic, real-time data is gathered through continuous glucose monitors that paint a detailed picture of metabolic response to different foods, wearable devices that track heart rate variability (HRV) as a measure of nervous system resilience, and sleep trackers that analyze sleep architecture. Advanced practitioners incorporate regular blood testing, measuring everything from lipid panels and hormone levels to inflammatory markers like CRP and vitamin D status. This constant flow of data creates a rich longitudinal dataset, moving the biohacker from a state of biological guesswork to one of empirical observation. The individual becomes both the experiment and the source of data, building a baseline against which any intervention can be measured.

10.2.2 Intervention Loop: From Data to Actionable Insight

Raw data, in isolation, is meaningless. The power of precision biohacking emerges from the rigorous interpretation of this data to inform highly specific interventions. This process creates a closed feedback loop: measure, intervene, and re-measure. For instance, a continuous glucose monitor might reveal that a seemingly healthy food, such as a bowl of oatmeal, causes a significant and prolonged spike in blood sugar for a particular individual, while a sweet potato does not.

The personalized intervention is clear: adjust breakfast choices accordingly to maintain metabolic stability. This loop becomes increasingly sophisticated as more variables are added. Correlating macronutrient intake with blood glucose levels, workout performance, and subjective mood scores can reveal optimal dietary patterns for that person. Tracking HRV alongside training load can pinpoint the precise point where exercise becomes counterproductive, signaling a need for recovery. The interventions themselves are equally personalized. Nutrigenomic insights might lead someone to supplement with methylated B vitamins due to an MTHFR gene mutation, while another individual might focus on optimizing omega-3 intake based on a specific fatty acid profile. This is a far cry from generic advice; it is a bespoke protocol, dynamically adjusted based on the ongoing stream of self-generated data. The biohacker acts as their own primary investigator, constantly formulating and testing hypotheses about what their unique biology needs to thrive.

10.2.3 Challenges of Hyper-Personalization: Signal vs. Noise

The pursuit of precision biohacking is not without its significant challenges and pitfalls. The most formidable obstacle is the problem of distinguishing true signal from noise within a vast sea of personal data. The human body is a complex, nonlinear system with countless confounding variables. A change in a biomarker could be the result of a specific intervention, or it could be caused by an unmeasured factor like a fleeting stressor, a minor illness, or even the weather.

Establishing causality in an N-of-1 experiment requires meticulous control of variables and replication of results, a process that demands immense discipline and can be prone to cognitive biases. The biohacker may see a pattern where none exists, a manifestation of the placebo effect or apophenia. Furthermore, the interpretation of complex data requires a high degree of scientific literacy. Misunderstanding a genetic report can lead to unnecessary anxiety or misguided interventions. There is also the risk of becoming overly obsessed with the metrics, a state sometimes termed "quantified self-itis," where the joy of living is subsumed by a constant anxiety to optimize every single data point. The goal of precision is not to become a slave to the data, but to use it as a guide to cultivate a deeper, more intuitive understanding of one's own body, ultimately using the tools to enhance rather than diminish personal freedom and well-being.

10.2.4 Future of the Individualized Model: AI and Systems Biology

The future of precision biohacking lies in overcoming these challenges through technology, specifically the application of artificial intelligence (AI) and systems biology. The complexity of the datasets being generated is already surpassing the analytical capabilities of the human brain. AI-powered platforms are emerging to fill this gap, capable of identifying subtle, non-obvious correlations across genomic, biomarker, and lifestyle data that a human might never detect.

These algorithms can move from correlation towards prediction, suggesting personalized interventions—a specific nutrient, a timing strategy for exercise, a sleep protocol—with a high probability of success for that individual. This transforms the biohacker from a lone analyst into a conductor, overseeing an AI co-pilot that handles the complex computational heavy lifting. Furthermore, the field is moving towards a more holistic, systems biology approach. Instead of looking at genes or biomarkers in isolation, the focus is on how they interact within complex networks. Precision biohacking will increasingly seek to understand how a dietary change influences gene expression (nutrigenomics), which in turn alters the gut microbiome (microbiomics), and how that subsequently affects immune function and mental state. This integrated, multi-omics approach will generate a truly holistic model of the individual, allowing for interventions that are not only personalized but also systemic, addressing root causes rather than isolated symptoms.

10.3 INTEGRATION WITH MAINSTREAM HEALTHCARE

10.3.1 Growing Convergence of Biohacking and Medicine

The integration of biohacking with mainstream healthcare reflects a broader shift toward personalized and preventive medicine. As patients become more proactive in managing their health, many are turning to biohacking tools such as wearable devices, genetic testing, and nootropic supplementation to optimize their well-being. This trend has prompted healthcare providers to acknowledge the value of patient-generated data and self-experimentation, particularly in areas where traditional medicine has limitations.

One of the most visible examples of this convergence is the use of continuous glucose monitors (CGMs) by non-diabetic individuals to optimize metabolic health. Originally designed for diabetes management, these devices are now being adopted by biohackers seeking to fine-tune their diets and exercise regimens. Similarly, direct-to-consumer genetic testing services like 23andMe have bridged the gap between personal curiosity and clinical genetics, allowing individuals to explore their genetic predispositions and discuss findings with healthcare providers.

This growing overlap suggests that biohacking is not merely an alternative to mainstream medicine but a potential extension of it. By embracing patient-driven health optimization, the medical field can benefit from the insights and innovations emerging from the biohacking community.

10.3.2 Wearable Technology and Self-Tracking in Clinical Settings

Wearable health technology has become one of the most significant points of intersection between biohacking and mainstream healthcare. Devices such as smartwatches, fitness trackers, and sleep monitors provide real-time biometric data that can inform both personal health decisions and clinical assessments. While biohackers have long used these tools to track and optimize their physiology, healthcare providers are increasingly recognizing their value in remote patient monitoring and chronic disease management.

For example, wearable ECG monitors can detect atrial fibrillation, prompting users to seek medical evaluation before symptoms worsen. Similarly, sleep-tracking devices can help identify patterns of insomnia or sleep apnea, leading to earlier interventions. The integration of these technologies into clinical practice allows for more continuous and personalized patient care, reducing the reliance on intermittent office visits and enabling proactive health management.

However, the challenge lies in ensuring the accuracy and reliability of consumer-grade wearables. While many devices provide useful insights, they are not always medically validated, which can lead to misinterpretation of data or unnecessary anxiety. As biohacking tools become more integrated into healthcare, there will be a growing need for standardized validation and physician oversight to ensure their safe and effective use.

10.3.3 Genetic Biohacking and the Future of Precision Medicine

Genetic biohacking represents one of the most controversial yet promising areas of integration between DIY biology and mainstream healthcare. Advances in CRISPR gene-editing technology have made it possible for individuals to explore genetic modifications, raising questions about how these practices might intersect with clinical genetics. While professional gene therapy remains tightly regulated, the accessibility of genetic testing and bioinformatics tools has empowered biohackers to experiment with personalized health strategies.

For instance, some biohackers use genetic data to tailor their diets, supplements, and exercise routines based on their unique genetic profiles. This approach aligns with the principles of precision medicine, where treatments are customized to an individual's genetic makeup. However, the lack of clinical oversight in genetic biohacking poses risks, such as misinterpretation of genetic data or the pursuit of unproven interventions.

The medical community is beginning to explore how patient-driven genetic experimentation might inform clinical practice. Some forward-thinking physicians are open to discussing genetic biohacking with patients, provided it is done responsibly and with proper guidance. This collaboration could lead to more personalized and effective healthcare strategies, but it also requires clear ethical boundaries to prevent misuse or harm.

10.3.4 Nootropics, Supplements, and Cognitive Enhancement

The use of nootropics and cognitive-enhancing supplements is another area where biohacking and mainstream healthcare are finding common ground. While traditional medicine has long prescribed stimulants like Adderall for ADHD, biohackers have expanded the use of cognitive enhancers to include off-label pharmaceuticals, herbal supplements, and experimental compounds. This trend has prompted discussions within the medical community about the potential benefits and risks of non-prescription cognitive enhancement. Some healthcare providers are now more open to discussing nootropic use with patients, particularly when it comes to well-researched supplements like omega-3 fatty acids, creatine, or even low-dose caffeine for cognitive performance. However, the lack of regulation in the supplement industry means that many products marketed as cognitive enhancers may lack scientific backing or contain harmful contaminants. This creates a need for greater collaboration between biohackers and medical professionals to ensure that cognitive enhancement is pursued safely and effectively.

10.3.5 Regulatory and Ethical Challenges

The integration of biohacking with mainstream healthcare is not without its challenges, particularly in terms of regulation and ethics. Traditional medicine operates under strict regulatory frameworks to ensure patient safety, whereas biohacking often exists in a legal gray area where oversight is minimal. This disparity raises concerns about the potential risks of unregulated self-experimentation and the ethical implications of patient-driven health optimization.

For example, while medical professionals are bound by ethical guidelines such as informed consent and non-maleficence, biohackers may not always adhere to these principles. This can lead to situations where individuals pursue risky modifications without fully understanding the consequences. Additionally, the commercialization of biohacking tools—such as at-home gene-editing kits or untested nootropics—poses regulatory challenges, as these products may not meet the same safety standards as medical treatments.

To address these concerns, there is a growing need for dialogue between the biohacking community and healthcare regulators. Establishing guidelines for safe experimentation, promoting transparency in self-tracking, and encouraging collaboration with medical professionals could help bridge the gap between these two approaches to health.

10.3.6 Role of Telemedicine and Digital Health Platforms

Telemedicine and digital health platforms are playing an increasingly important role in integrating biohacking with mainstream healthcare. These technologies allow patients to share self-tracked data with healthcare providers, facilitating more personalized and continuous care. For example, a patient using a wearable device to monitor heart rate variability might share this data with their cardiologist to adjust treatment plans in real time.

Similarly, digital health platforms that combine biohacking tools with clinical oversight are emerging as a new model of healthcare delivery. These platforms can provide patients with actionable insights while ensuring that their self-experimentation is guided by medical expertise. This hybrid approach could lead to more efficient and patient-centered healthcare, where individuals are empowered to take an active role in their well-being without sacrificing safety or efficacy.

10.4 CORPORATE BIOHACKING PROGRAMS AND PERFORMANCE LABS

10.4.1 Evolution of Corporate Wellness to Biohacking

Corporate wellness programs have existed for decades, typically focusing on basic initiatives such as fitness memberships, smoking cessation, and health screenings. While effective to a certain extent, these programs often lacked personalization and relied on generalized approaches to health. With the rise of biohacking, companies began to see opportunities to go beyond traditional wellness initiatives by offering employees access to advanced tools, data-driven strategies, and personalized interventions. The shift from wellness to biohacking reflects a broader cultural change, where employees are not only expected to maintain basic health but also encouraged to pursue peak performance and longevity. By embracing biohacking, corporations acknowledge that performance is deeply tied to biology and that optimizing physiology and cognition can lead to measurable business outcomes.

10.4.2 Performance Labs

Performance labs represent the physical and conceptual spaces where corporate biohacking programs are developed, tested, and implemented. These labs are designed to provide employees with access to cutting-edge technologies and controlled environments where biological performance can be measured and improved. For instance, performance labs may include biometric testing stations that measure heart rate variability, brainwave monitoring devices, or metabolic analysis tools. They often integrate recovery pods, cryotherapy chambers, or sleep optimization setups, creating environments that allow individuals to experiment with new approaches to resilience and energy management. These spaces are not limited to elite athletes or military professionals but are increasingly available to corporate employees, reflecting the growing recognition that workplace performance can benefit from scientific and technological innovations traditionally reserved for specialized fields.

10.4.3 Biohacking Practices in Corporate Programs

The practices adopted in corporate biohacking programs are varied, reflecting the diversity of the biohacking movement itself. Many programs focus on cognitive enhancement through interventions such as nootropic supplementation, neurofeedback training, or mindfulness supported by brain-computer interface devices. Others prioritize physical performance by offering strategies for metabolic flexibility, personalized nutrition, and sleep optimization. Some corporations even integrate light therapy, grounding techniques, and circadian rhythm synchronization protocols into their performance labs to ensure employees maintain optimal energy throughout the day. By blending physical, cognitive, and emotional optimization, corporate biohacking programs create comprehensive models for employee performance that surpass the traditional view of workplace wellness.

10.4.4 Role of Data and Personalization

Central to corporate biohacking is the collection and analysis of biometric and behavioral data. Employees often undergo detailed testing, from genetic screenings to continuous glucose monitoring, enabling personalized recommendations for diet, exercise, and recovery. Artificial intelligence platforms are increasingly employed to analyze these datasets and generate dynamic optimization plans. This personalization ensures that interventions are not generic but tailored to the unique biological profile of each individual. For corporations, this not only improves health outcomes but also demonstrates a commitment to valuing employees as individuals. However, the heavy reliance on personal data also raises concerns about privacy and ethical use, highlighting the need for clear policies on how information is collected, stored, and applied in workplace contexts.

10.4.5 Enhancing Productivity and Innovation

Corporate biohacking programs are not solely focused on health for its own sake but are often implemented with the explicit goal of enhancing productivity and innovation. The rationale is that healthier, more energized, and more focused employees will naturally contribute to better organizational performance. Companies that adopt biohacking practices often report improvements in concentration, faster problem-solving, and greater resilience to stress. Performance labs provide employees with the tools to recover quickly from fatigue, maintain high energy levels, and sustain creative thinking. By investing in these initiatives, corporations aim to cultivate environments where employees operate closer to their biological and cognitive potential, driving overall innovation within the company.

10.4.6 Ethical and Cultural Considerations

While the benefits of corporate biohacking programs are appealing, they also raise important ethical and cultural questions. Employees may feel pressured to participate in programs that involve invasive testing or experimental interventions, blurring the line between voluntary participation and workplace obligation. There are also concerns about data security, as sensitive genetic or biometric information could be misused or mishandled. From a cultural perspective, corporate biohacking programs can reflect a double-edged sword: on one hand, they demonstrate care for employee well-being, while on the other, they risk creating environments where health optimization becomes another metric of performance. Balancing empowerment with ethical safeguards is therefore essential in ensuring that biohacking in corporate contexts enhances employee lives rather than placing new burdens upon them.

10.4.7 Global Expansion of Performance Labs

Corporate biohacking programs and performance labs are not limited to Silicon Valley or Western corporations but are increasingly appearing in global markets. Technology companies, financial institutions, and even government agencies are experimenting with these initiatives as part of their talent management strategies. In Asia, where cultural emphasis on collective well-being intersects with rapid technological adoption, biohacking practices are being integrated into large corporate ecosystems. In Europe, ethical considerations and sustainability are shaping the development of performance labs that prioritize holistic health and long-term resilience. This global spread demonstrates that corporate biohacking is not a passing trend but a movement gaining traction across industries and cultures.

10.5 TOWARD HUMAN ENHANCEMENT AND TRANSHUMANISM

10.5.1 Idea of Human Enhancement

Human enhancement refers to deliberate interventions designed to improve physical, cognitive, or emotional capacities beyond what is considered normal or natural. This may include increasing strength and endurance, sharpening mental focus, extending lifespan, or even adding entirely new abilities such as expanded sensory perception. In the context of biohacking, human enhancement often begins with small-scale interventions like nutritional optimization, wearable devices, or self-experimentation with nootropics. However, as the movement progresses, more advanced technologies such as genetic engineering, brain-computer interfaces, and synthetic biology are increasingly explored. These interventions blur the distinction between therapy, which seeks to restore normal function, and enhancement, which aims to push human capacity beyond its natural limits.

10.5.2 Biohacking as a Pathway to Enhancement

Biohacking plays a unique role in advancing human enhancement because it democratizes access to tools and knowledge. Unlike traditional biomedical research, which is often confined to academic institutions and corporate laboratories, biohacking operates through community labs, open-source platforms, and personal experimentation. This grassroots approach accelerates innovation and allows individuals to test new methods of self-improvement without waiting for institutional approval. For example, biohackers have experimented with implantable chips for digital identification, genetic editing tools like CRISPR for self-directed modification, and continuous monitoring devices that provide real-time feedback on physiology. While not all experiments are successful or safe, they reflect a growing cultural willingness to view the human body as a system that can be optimized, reprogrammed, and upgraded.

10.5.3 Transhumanism and Its Philosophical Foundations

Transhumanism extends the ambitions of biohacking into a comprehensive worldview that advocates for the deliberate use of technology to transcend human limitations. Rooted in Enlightenment ideals of progress and reason, transhumanism envisions a post-human future where aging, disease, and even death can be overcome. Central to transhumanism is the belief that human beings should not accept biological constraints as fixed but instead strive to enhance their capabilities through scientific and technological means.

This philosophy inspires biohackers who see their work not just as self-improvement but as part of a larger project to redefine the trajectory of human evolution. While some critics argue that transhumanism risks undermining the essence of humanity, proponents view it as the natural continuation of humanity's long history of innovation.

10.5.4 Technologies Driving Enhancement

The pursuit of human enhancement through biohacking and transhumanism is powered by a range of emerging technologies. Artificial intelligence plays a crucial role in processing genetic and physiological data to create personalized interventions. Brain-computer interfaces allow direct communication between the mind and machines, potentially enabling cognitive augmentation and new forms of interaction. Synthetic biology makes it possible to design organisms or modify genetic codes to achieve desirable traits.

Nanotechnology promises microscopic interventions in cells and tissues that could repair damage or enhance performance. Longevity science explores genetic and molecular pathways to slow aging and extend human lifespan. These technologies, once confined to science fiction, are increasingly being tested by biohackers and supported by transhumanist thinkers as tangible steps toward a post-human future.

10.5.5 Pursuit of Immortality

One of the most ambitious and controversial aspects of transhumanism is the pursuit of immortality or radical life extension. Biohacking communities are deeply engaged in longevity research, experimenting with anti-aging compounds, caloric restriction, and regenerative therapies. Some projects explore cryonics, the preservation of bodies or brains for potential future revival, while others focus on mind uploading, the theoretical transfer of human consciousness into digital systems. While many of these ideas remain speculative, they reflect a broader desire to overcome mortality and redefine human existence. Biohacking, with its emphasis on experimentation and personal agency, serves as a testing ground for these ideas, making the pursuit of extended life more accessible to individuals outside traditional scientific frameworks.

10.5.6 Ethical Challenges of Human Enhancement

The movement toward human enhancement and transhumanism raises profound ethical questions. Issues of equity and accessibility are among the most pressing, as advanced enhancements may be available only to wealthy individuals, creating a divide between enhanced and non-enhanced populations. Questions of identity also emerge: if human beings significantly alter their biology or merge with machines, what remains of their humanity? There are concerns about unintended consequences, such as genetic modifications that introduce unforeseen health problems or technologies that disrupt social cohesion. Additionally, the pursuit of immortality raises philosophical dilemmas about overpopulation, environmental sustainability, and the meaning of life itself. While biohacking encourages personal freedom and experimentation, it also highlights the need for ethical frameworks that guide the responsible use of enhancement technologies.

10.5.7 Cultural and Social Implications

Beyond the individual level, human enhancement through biohacking has profound cultural and social implications. It challenges traditional notions of health, ability, and normality, forcing societies to reconsider what counts as an acceptable or desirable human trait. In some cultures, enhancement may be embraced as progress, while in others it may be viewed as unnatural or threatening to human dignity. The adoption of enhancements in workplaces, education, or sports also raises concerns about fairness and competition. If cognitive or physical augmentations become widespread, those without access could be disadvantaged, leading to new forms of inequality. On the other hand, if implemented inclusively, enhancements could alleviate suffering, eliminate disease, and empower individuals to achieve greater potential.

CHAPTER 11

Case Studies and Self-Experimentation Logs

11.1 DOCUMENTED EXPERIMENTS AND PROTOCOLS

11.1.1 Importance of Documentation in Biohacking

Documentation is central to biohacking because it allows experiments to move from personal anecdotes to structured evidence. Many biohackers engage in self-experimentation, but without proper records, their experiences remain subjective and non-transferable. When protocols are written and experiments are documented with detail, they can be repeated, evaluated, and improved by others. This practice mirrors the scientific method and helps establish credibility in a field often criticized for its lack of rigor. Documentation also serves as a protective measure, reducing risks by outlining steps clearly and making it easier to identify potential errors. In community labs, shared records create transparency and encourage collaboration, enabling citizen scientists to build on each other's work.

11.1.2 Categories of Documented Biohacking Experiments

Documented biohacking experiments fall into several categories that reflect the diversity of the movement. Nutritional and metabolic biohacking experiments often involve testing the effects of specific diets, supplements, or fasting protocols on biomarkers such as glucose levels, heart rate variability, or cognitive performance. Cognitive and neuroenhancement experiments explore interventions ranging from meditation and brainwave entrainment to nootropics and neurofeedback training. Genetic biohacking protocols focus on the use of tools like CRISPR to modify DNA in microorganisms or, in some cases, in human cells. Other experiments explore environmental and sensory biohacking, including the impact of light exposure, grounding, or temperature stress on physiology. Each category relies on protocols that outline variables, methods of measurement, and desired outcomes, reflecting a move toward systematic inquiry.

11.1.3 Case Studies of Documented Experiments

Several well-documented biohacking experiments illustrate both the potential and the challenges of the movement. In the realm of nutrition, many biohackers have shared detailed accounts of ketogenic diet protocols, documenting changes in weight, energy levels, and blood markers over extended periods. In cognitive enhancement, experiments involving nootropics have been meticulously recorded, with individuals tracking variables such as memory recall, focus, and mood under controlled supplementation schedules. More radical examples include experiments where biohackers attempted to use CRISPR kits to edit their own DNA, documenting step-by-step procedures and results. These cases often spark debates about safety and ethics but also demonstrate the willingness of biohackers to apply rigorous self-tracking to test unconventional ideas. Documented experiments of this nature provide insight into both successes and failures, offering a valuable resource for others who may wish to replicate or adapt the protocols.

11.1.4 Protocols in Community Biohacking Labs

Community laboratories are at the forefront of establishing formal biohacking protocols. These labs operate much like academic research centers but with open access to anyone interested in exploring biology. In such environments, documentation takes the form of shared lab notebooks, online repositories, and collaborative platforms where protocols are standardized and results are published for public use. For instance, protocols for bacterial gene editing or yeast fermentation are often openly shared, enabling amateur scientists to follow precise steps while contributing their own modifications. These community-driven documentation practices not only ensure reproducibility but also democratize science, allowing individuals outside traditional institutions to engage in meaningful biological research.

11.1.5 Role of Self-Tracking Protocols

Self-tracking is one of the most common forms of biohacking, and it relies heavily on documented protocols. Biohackers often design experiments where they monitor physiological data such as sleep cycles, blood glucose, or brain activity over time while introducing controlled variables like dietary changes, supplements, or environmental shifts. Protocols for self-tracking emphasize consistency and precision, outlining how measurements should be taken, what tools should be used, and how results should be recorded. These records are often shared through blogs, online forums, or social media platforms, creating a collective knowledge base. The rigor of these protocols determines the quality of the insights gained, highlighting the importance of careful documentation in transforming personal experiences into community knowledge.

11.1.6 Ethical Considerations in Documented Experiments

The documentation of biohacking experiments brings with it significant ethical considerations. Transparency ensures that others can learn from past experiments, but it also raises concerns about safety and misuse. Detailed protocols involving genetic modification, for example, could be replicated by individuals without adequate safety precautions, creating risks for both experimenters and the environment. Ethical documentation must balance openness with responsibility, ensuring that potentially dangerous methods are contextualized with warnings and biosafety guidelines. Additionally, the documentation of personal data, such as genetic information or health metrics, raises questions about privacy and informed consent, especially when records are shared publicly. Ethical frameworks in biohacking documentation are still evolving, but they are increasingly recognized as necessary for the movement's credibility.

11.1.7 Influence of Documented Experiments on Mainstream Science

Documented biohacking experiments, while often informal, can have significant influence on mainstream science and medicine. In some cases, biohacker-led experiments have provided early insights that inspired more formal research studies. For example, self-tracking protocols around sleep optimization and circadian rhythm synchronization have informed professional research into sleep science. Similarly, nootropic experiments conducted by biohacking communities have influenced studies into cognitive enhancement. By providing anecdotal but structured data, biohackers create hypotheses that can later be tested in controlled settings. This interplay highlights how biohacking and traditional science are not mutually exclusive but can complement one another when documentation is rigorous and transparent.

11.1.8 Challenges

One of the biggest challenges in biohacking is the standardization of protocols. Unlike academic research, which follows strict methodologies and peer review processes, biohacking is decentralized and highly variable. Different individuals may apply different measurement tools, record data in inconsistent formats, or overlook key variables. This variability makes it difficult to replicate experiments reliably. Efforts are underway in many community labs and online platforms to create standardized templates for documenting protocols, ensuring that methods are consistent and results are comparable. The development of these shared standards is crucial for advancing biohacking from a collection of isolated projects to a recognized field of experimental practice.

11.2 SUCCESS STORIES AND TRANSFORMATIONS

The journey into biohacking often begins with a catalyst—a moment of profound dissatisfaction with one's current state or a burning desire to achieve a level of performance previously thought unattainable. For many, this manifests as a struggle with chronic fatigue, brain fog, stubborn weight, or a general sense of being unwell despite conventional medicine finding no overt disease. This was the case for countless individuals who turned to biohacking as a last resort. They began with the foundational hacks: meticulous tracking of their nutrition, experimenting with the elimination of inflammatory foods like gluten and sugar, and adopting dietary frameworks such as ketosis or intermittent fasting. The transformations reported are often dramatic. Individuals speak of shedding dozens of pounds not through sheer willpower, but by understanding the hormonal responses of their body to different macronutrients. They describe the dissipation of mental fog, replaced by a sharp, sustained clarity and energy levels they had not experienced since youth. This nutritional biohacking is not about fad diets; it is a systematic process of using blood glucose monitors and food journals to create a personalized eating protocol that aligns with their unique metabolic signature.

Beyond nutrition, the realm of sleep optimization showcases some of the most immediate and impactful biohacking successes. For generations, poor sleep was often accepted as an unfortunate byproduct of modern life. Biohackers refuse this notion. By employing technology like wearable sleep trackers such as the Oura Ring or Whoop strap, individuals gain unprecedented insight into their sleep architecture.

They can see the precise duration of their deep sleep and REM cycles, track their resting heart rate variability (HRV)—a key marker of recovery and stress—and monitor their ambient bedroom temperature and light exposure. Armed with this data, they experiment. They implement strict digital curfews to block blue light, regulate room temperature to the optimal 65-68 degrees Fahrenheit, and utilize supplements like magnesium to enhance sleep quality. The results are transformative. Executives report making better decisions under pressure. Athletes note significantly faster recovery times and improved performance metrics. Artists and writers describe enhanced creativity and cognitive fluency. By hacking sleep, they are not just resting; they are fundamentally upgrading their brain's maintenance and consolidation processes every single night.

The pursuit of cognitive enhancement, or "neurohacking," is another area rich with success stories. In a world demanding constant focus and innovation, the ability to sharpen one's mind is a prized advantage. Biohackers in this domain employ a diverse toolkit. Nootropic stacks—carefully formulated combinations of natural and synthetic compounds like lion's mane mushroom, caffeine + L-theanine, and creatine—are used to boost memory, focus, and neuroplasticity. Technologies like transcranial direct current stimulation (tDCS) devices are experimented with to modulate neuronal activity. Perhaps the most powerful tool, however, is the practice of mindfulness and meditation, quantified through neurofeedback. Users wear EEG headbands that provide real-time data on their brainwave states, allowing them to literally see when their mind is calm and focused and when it is distracted. This biofeedback loop accelerates the mastery of meditation, leading to profound reductions in anxiety and stress. Individuals once crippled by chronic anxiety describe learning to control their physiological stress response, transforming their mental landscape from one of reactivity to one of calm agency. This is a quintessential biohacking transformation: using data to gain mastery over the once-abstract realm of emotion and thought.

Perhaps the most visually dramatic and emotionally resonant biohacking transformations occur in the field of longevity and anti-aging. This is where biohacking transcends mere optimization and enters the domain of life extension. Pioneers in this space act as their own test subjects, employing a rigorous regime of biomarker testing. They regularly analyze their blood for everything from standard cholesterol panels to more advanced markers like apolipoprotein B, insulin-like growth factor 1 (IGF-1), and inflammatory cytokines.

They use DNA testing from companies like 23andMe to understand their genetic predispositions and then craft personalized interventions to mitigate those risks. Their protocols can be intense, involving cold plunge therapy to reduce inflammation and boost dopamine, hyperbaric oxygen chambers to enhance cellular repair, and sophisticated supplementation regimens like NMN or resveratrol to support cellular health. While the long-term goal of significantly extending human lifespan is still a subject of research, the short-term successes are undeniable. These biohackers often boast biological ages—calculated based on biomarkers—that are ten to twenty years younger than their chronological age. Their transformation is one of vitality, possessing the energy, physique, and cellular health of a much younger person, effectively compressing morbidity and expanding their healthspan.

It is crucial to acknowledge that the path of biohacking is not without its perils. The ethos of self-experimentation requires a foundation of knowledge, a respect for the scientific method, and, ideally, guidance from qualified medical professionals. The most successful biohacking stories are not about reckless abandon but about the methodical, safe, and incremental application of interventions while carefully tracking outcomes.

The transformation, therefore, is not just physical or cognitive; it is also philosophical. Successful biohackers undergo a fundamental shift in identity. They cease to be passengers in the vehicle of their body and become its pilot, engineer, and navigator. They develop a deep sense of agency and ownership over their health, understanding that their daily choices collectively determine their long-term trajectory. They move from a model of treating disease to one of optimizing wellness, from reacting to illness to proactively building resilience. In these countless success stories, from the executive who hacked her sleep to defeat burnout to the retiree who hacked his biomarkers to regain the vitality of his forties, we see the emergence of a new human archetype: one empowered by data, driven by curiosity, and dedicated to the endless project of forging their own best self.

11.3 BIOHACKING IN CHRONIC DISEASE MANAGEMENT

11.3.1 Rise of Patient-Driven Chronic Disease Management

The traditional model of chronic disease management has often been reactive, focusing on symptom control rather than prevention or root-cause resolution. Biohacking introduces a more proactive and personalized approach, where patients become active participants in their own care. This shift is being driven by several factors, including the increasing availability of health-tracking technologies, the growing body of research on lifestyle interventions, and a cultural movement toward self-optimization.

Patients with chronic conditions are turning to biohacking as a means of gaining better control over their health outcomes. For example, individuals with type 2 diabetes are using continuous glucose monitors not just for medical management but to experiment with dietary interventions that might improve their metabolic health. Similarly, those with autoimmune conditions are exploring personalized nutrition plans, stress-reduction techniques, and targeted supplementation based on genetic testing and biomarker analysis. This patient-driven approach represents a fundamental change in how chronic diseases are managed, moving from a one-size-fits-all model to one that is tailored to individual biology and lifestyle.

11.3.2 Wearable Technology and Continuous Health Monitoring

One of the most significant contributions of biohacking to chronic disease management has been the adoption of wearable health technologies. These devices provide real-time data that can help patients and healthcare providers make more informed decisions about treatment and lifestyle adjustments. For conditions that require constant monitoring, such as diabetes or hypertension, wearables offer a level of insight that was previously unattainable outside clinical settings. Continuous glucose monitors (CGMs) have become particularly valuable for individuals with diabetes, allowing them to track how different foods, exercise routines, and stress levels affect their blood sugar. This data can be used to refine dietary choices, optimize medication timing, and identify patterns that might otherwise go unnoticed. Similarly, wearable ECG monitors can help those with cardiovascular conditions detect arrhythmias early, while sleep trackers can provide insights into how rest quality affects chronic pain or autoimmune flare-ups. The key advantage of these technologies is their ability to provide actionable, personalized data. Rather than relying solely on periodic doctor visits, patients can now monitor their conditions continuously and make adjustments in real time. This shift toward data-driven self-management represents a significant advancement in chronic disease care, though it also requires patients to develop the skills to interpret and act on this information responsibly.

11.3.3 Nutritional Biohacking for Chronic Conditions

Nutrition plays a crucial role in managing chronic diseases, and biohacking approaches have introduced new ways to optimize dietary interventions. Traditional dietary advice for chronic conditions often follows general guidelines, but biohacking encourages a more personalized approach based on individual biomarkers, genetic predispositions, and real-time physiological responses. This tailored nutrition strategy can be particularly effective for conditions influenced by diet, such as metabolic syndrome, inflammatory bowel disease, and certain autoimmune disorders.

For example, some biohackers with chronic conditions use at-home testing kits to monitor biomarkers like inflammation, gut microbiome composition, or nutrient deficiencies. This data can then inform highly personalized dietary plans that go beyond standard recommendations. Others experiment with fasting protocols, ketogenic diets, or targeted supplementation to determine what works best for their specific condition and biology. While these approaches can yield significant benefits, they also require careful monitoring to avoid potential nutritional imbalances or adverse reactions.

The challenge with nutritional biohacking lies in distinguishing between evidence-based interventions and unproven trends. As patients increasingly take dietary management into their own hands, there is a growing need for healthcare providers to engage with these practices, helping to guide patients toward safe and effective nutritional strategies.

11.3.4 Genetic Testing and Personalized Medicine

Advances in genetic testing have opened new possibilities for personalized chronic disease management through biohacking approaches. Direct-to-consumer genetic tests provide individuals with insights into their predispositions for certain conditions, their likely responses to medications, and even their optimal dietary and exercise strategies based on genetic markers. This information can be particularly valuable for managing chronic diseases with strong genetic components, such as certain types of diabetes, cardiovascular diseases, and some autoimmune conditions.

For instance, individuals with a genetic predisposition to high cholesterol might use this information to adopt more aggressive lifestyle interventions earlier in life. Those with genetic variants affecting nutrient metabolism might tailor their diets to compensate for these differences. While genetic testing provides powerful insights, it also requires careful interpretation to avoid misapplication or unnecessary anxiety about genetic risks.

The integration of genetic data with other biohacking tools creates opportunities for truly personalized chronic disease management. However, this approach also highlights the need for better education about genetic information and more collaboration between patients and healthcare providers to ensure that genetic insights are used responsibly and effectively.

11.3.5 Lifestyle Optimization and Behavioral Biohacking

Beyond technological and nutritional interventions, biohacking approaches to chronic disease management often include lifestyle and behavioral optimization strategies. These can range from sleep optimization and stress reduction techniques to exercise biohacking and environmental modifications. For many chronic conditions, these lifestyle factors play a crucial role in disease progression and symptom management. For example, individuals with chronic pain conditions might use biofeedback devices to learn stress-reduction techniques that help manage their symptoms. Those with cardiovascular diseases might experiment with different exercise protocols, using wearable data to determine the optimal intensity and duration for their specific condition. Sleep optimization is another area where biohacking can make a significant difference, as poor sleep is often both a consequence and a contributor to many chronic diseases.

The advantage of these lifestyle approaches is that they are generally low-risk and can be highly effective when properly implemented. However, they also require consistent effort and often benefit from professional guidance to ensure they are appropriately tailored to individual needs and medical conditions.

11.3.6 Challenges and Risks in Biohacking for Chronic Diseases

While biohacking offers exciting possibilities for chronic disease management, it is not without significant challenges and potential risks. One of the primary concerns is the lack of standardized safety protocols in many biohacking practices. Unlike medical treatments that undergo rigorous testing, many biohacking interventions are experimental and may lack sufficient evidence of long-term safety and efficacy.

There is also the risk of information overload and misinterpretation. With the abundance of health data available through biohacking tools, patients may struggle to determine which information is relevant and how to act on it appropriately. This can lead to unnecessary anxiety, inappropriate interventions, or even harmful self-experimentation. Additionally, the commercialization of biohacking tools and supplements creates opportunities for exploitation, where unproven or unsafe products might be marketed to vulnerable individuals seeking solutions for their chronic conditions. Perhaps the most significant challenge is the potential for biohacking to create a divide between patients and healthcare providers. If individuals pursue self-management strategies without medical oversight, they may miss important clinical insights or delay necessary treatments. This highlights the need for better integration between biohacking approaches and conventional medical care, where patients and providers work together to optimize health outcomes.

11.4 FAILURES, LEARNINGS, AND UNEXPECTED OUTCOMES

11.4.1 Inevitability of Failure in Experimental Practices

Failure is inherent in biohacking because the movement thrives on experimentation outside the rigid frameworks of conventional science. Biohackers often pursue interventions with incomplete knowledge, relying on emerging research or anecdotal reports. This creates room for error, ranging from ineffective results to health risks. For instance, attempts at cognitive enhancement with untested nootropics sometimes lead to side effects such as insomnia, anxiety, or digestive problems rather than the intended mental clarity. Similarly, fasting regimens or extreme diets may cause nutrient imbalances when not carefully managed. These failures underscore the reality that the body is a complex system, and tinkering with it is not always predictable. Rather than discouraging experimentation, such failures often motivate biohackers to refine their methods and share insights with the wider community.

11.4.2 Learning through Trial and Error

One of the defining features of biohacking is the emphasis on learning through trial and error. Failures are rarely seen as definitive setbacks but as opportunities to understand biological systems more deeply. When a protocol does not yield expected outcomes, biohackers examine their data, track variables, and attempt to identify where assumptions were flawed.

This iterative approach mirrors the scientific process, though it is often conducted in informal or personal settings. For example, a biohacker experimenting with sleep optimization may find that blue light filters or supplements like melatonin do not produce immediate results. Instead of abandoning the project, the individual may adjust environmental factors, alter dosage, or explore new tools, gradually building a more nuanced understanding of circadian rhythms. These learnings contribute not only to personal growth but also to the broader biohacking community when shared publicly.

11.4.3 Unexpected Outcomes as Catalysts for Innovation

Some of the most valuable contributions of biohacking arise not from carefully designed successes but from unexpected outcomes. In many cases, experiments aimed at one goal reveal entirely different effects, sparking new directions for exploration.

For instance, individuals experimenting with ketogenic diets for weight loss sometimes report enhanced mental clarity and focus, leading to further inquiry into the relationship between ketones and brain performance. Similarly, cryotherapy initially explored for recovery and inflammation management has been associated with mood improvement in certain individuals, prompting new research into the connection between cold exposure and mental health. These serendipitous discoveries highlight the creativity and openness that characterize biohacking, where surprises are not seen as errors but as invitations to expand inquiry.

11.4.4 Ethical Dilemmas Emerging from Failures

Failures and unexpected outcomes also raise important ethical questions within biohacking. Self-experimentation is often celebrated as a form of autonomy, but when results include harm, it becomes necessary to consider where boundaries should be drawn. Experiments involving unregulated genetic modification or extreme chemical supplementation can lead to irreversible consequences.

Documented cases of biohackers attempting DIY gene editing have raised concerns about the risks of encouraging replication by others who may not understand the dangers involved. Ethical dilemmas arise not only in terms of individual safety but also in how failures are communicated to the broader community. Sharing failures responsibly means providing context, warnings, and guidelines so others do not repeat the same mistakes without understanding the potential consequences.

11.4.5 Community Learnings from Documented Failures

One of the strengths of the biohacking movement lies in its culture of openness and community learning. Failures and mistakes are often documented in online forums, blogs, and community labs, creating valuable repositories of knowledge. Instead of concealing missteps, many biohackers share their experiences transparently, detailing what went wrong, what was learned, and how protocols might be improved.

This collective documentation transforms individual failures into communal wisdom, accelerating progress for everyone involved. For example, shared accounts of adverse effects from certain nootropic combinations have prevented others from repeating potentially harmful experiments. The willingness to disclose and discuss failures contributes to the resilience of the biohacking ecosystem, making it more robust than if only successes were recorded.

11.4.6 Psychological Resilience in the Face of Failure

Biohacking often requires psychological resilience, as failures can be personally challenging. Self-experimentation involves high expectations for transformation, whether in terms of enhanced cognition, improved fitness, or longevity. When results fall short, individuals may feel discouraged or even question their ability to influence biological systems at all.

However, many biohackers embrace a mindset that values persistence and adaptability. Psychological resilience allows them to see failures as temporary obstacles rather than final judgments. This mental framing is essential because it keeps experimentation alive and encourages a culture where exploration is more important than perfection. Resilience also empowers biohackers to continue experimenting responsibly, turning personal setbacks into stepping stones for future success.

11.4.7 Role of Technology in Identifying Failures

Modern biohacking increasingly relies on wearable devices and digital platforms to track data, which plays a critical role in identifying failures and unexpected outcomes. Devices that monitor heart rate variability, blood glucose, or sleep cycles provide objective feedback that helps biohackers assess whether interventions are working as intended. In some cases, technology reveals that perceived improvements are not supported by measurable data, leading to the realization that an experiment has failed.

Conversely, technology can also highlight unexpected outcomes, such as discovering that a practice designed for physical performance improvement may also benefit stress regulation. By making failures visible and quantifiable, technology enhances the learning process and ensures that experimentation is guided by evidence rather than assumption alone.

11.4.8 Failures as a Bridge to Mainstream Science

Interestingly, the failures and unexpected outcomes of biohacking often capture the attention of mainstream scientific communities. Citizen science experiments that do not succeed can inspire formal research by raising questions about why certain interventions fail or produce surprising effects. This dynamic creates a feedback loop where biohacking informs professional science and vice versa. Failures in biohacking, when documented and shared, can highlight gaps in existing scientific knowledge, prompting new investigations. For example, the inconsistent results of self-administered nootropic experiments have led researchers to reexamine how variables like genetics and environment influence cognitive performance. In this way, failures serve not only as internal lessons for biohackers but also as valuable inputs for the broader pursuit of scientific understanding.

11.4.9 Embracing Failure as a Philosophy

Within biohacking, failure is not simply tolerated but often embraced as a philosophy of growth and innovation. The unpredictability of human biology ensures that no experiment is guaranteed to succeed, but the willingness to accept this uncertainty opens doors to discovery. By reframing failures as feedback rather than defeat, biohackers cultivate a culture of continuous learning. This philosophy is deeply aligned with the ethos of experimentation, where each misstep is part of a larger journey toward understanding. Embracing failure helps biohackers remain adaptable, creative, and open to surprises, reinforcing the idea that progress is rarely linear but often shaped by detours and unanticipated insights.

CHAPTER 12

Appendices

12.1 GLOSSARY OF BIOHACKING TERMS

- **Adaptogens:** Natural substances, often herbs or mushrooms, believed to help the body adapt to stress and restore balance.
- **Artificial Intelligence (AI) in Biohacking:** The use of machine learning and AI tools to analyze biometric data and personalize biohacking interventions.
- **BCI (Brain-Computer Interface):** A direct communication pathway between the brain and an external device, used in advanced biohacking for control and enhancement.
- **Biofeedback:** A technique that uses monitoring devices to provide real-time data about physiological functions, enabling conscious regulation.
- **Biohacking:** The practice of applying science, technology, and self-experimentation to improve health, performance, and longevity.
- **Biometric Data:** Quantitative information about biological functions, such as heart rate, blood glucose, or sleep cycles, often tracked in biohacking.
- **Citizen Science:** Scientific research carried out by non-professionals, often forming the foundation of community-based biohacking.
- **Cold Exposure:** A biohacking practice involving deliberate exposure to cold, such as ice baths or cryotherapy, to build resilience and improve circulation.
- **CRISPR:** A gene-editing tool used in genetic biohacking for precise modification of DNA.
- **Cryotherapy:** A treatment involving extremely cold temperatures to reduce inflammation and enhance recovery.
- **DIYbio (Do-It-Yourself Biology):** A grassroots movement where individuals and groups conduct biological experiments outside formal labs.

- **DNA Data Storage**: The experimental practice of encoding digital information within DNA strands, part of molecular biohacking.
- **Epigenetics**: The study of changes in gene expression influenced by environment and lifestyle rather than DNA sequence changes.
- **Fasting (Intermittent Fasting)**: A dietary strategy where eating is restricted to specific time windows, used in biohacking for metabolic and cognitive benefits.
- **Grounding (Earthing)**: Direct physical contact with the earth, such as walking barefoot, believed to balance energy and reduce stress.
- **Implantables**: Devices implanted under the skin, such as RFID chips, used by some biohackers to extend capabilities.
- **Ketosis**: A metabolic state where the body burns fat instead of glucose, often induced by ketogenic diets for energy and mental clarity.
- **Longevity Science**: Biohacking focused on extending lifespan and healthspan through interventions like supplements, diet, and genetic research.
- **Microbiome**: The collection of microorganisms living in the human body, particularly the gut, targeted in biohacking to improve health.
- **Nootropics**: Natural or synthetic compounds used to enhance cognitive function, memory, focus, and creativity.
- **Performance Labs**: Specialized facilities for testing biohacking interventions on productivity, endurance, and cognitive performance.
- **Quantified Self**: A movement in which individuals track personal health data using devices and apps to optimize performance and well-being.
- **Red Light Therapy**: The use of low-level wavelengths of red or near-infrared light to promote cellular repair and reduce inflammation.
- **Resilience Training**: Biohacking techniques aimed at improving mental and physical adaptability under stress.
- **Synthetic Biology**: The design and construction of new biological systems for innovation and enhancement, a cutting-edge area of biohacking.
- **Transhumanism**: A philosophy within biohacking that advocates using technology to surpass human biological limits.
- **Wearables**: Devices such as smartwatches, glucose monitors, or EEG headbands that track biometric data for biohacking experiments.

12.2 LIST OF TOOLS, DEVICES, AND VENDORS

12.2.1 Wearable Devices

- **Oura Ring:** A smart ring that tracks sleep, recovery, activity, and readiness metrics.
- **Whoop Strap:** A band that monitors strain, recovery, and sleep, popular among athletes.
- **Apple Watch:** Multi-functional smartwatch with heart rate, ECG, and fitness tracking.
- **Fitbit:** Fitness and health tracker with stress monitoring and activity logging.
- **Biostrap:** Advanced biometrics wearable focused on heart rate variability (HRV) and sleep.

12.2.2 Genetic and Molecular Tools

- **23andMe:** Consumer genetic testing service for ancestry and health traits.
- **Nebula Genomics:** Whole-genome sequencing for deep genetic insights.
- **SelfDecode:** Personalized health recommendations based on genetic data.
- **CRISPR Kits (Amino Labs, The ODIN):** DIY kits for learning gene-editing techniques.

12.2.3 Brain and Cognitive Enhancement Devices

- **Muse Headband:** EEG-based device for meditation and brain activity monitoring.
- **Halo Neuroscience:** Neurostimulation headset designed to enhance learning and focus.
- **FocusCalm:** EEG headband for relaxation and mental performance.
- **Neurable:** Brain-computer interface (BCI) headset enabling neural interaction.

12.2.4 Light and Sensory Tools

- **Joovv:** Red light and near-infrared therapy panels for recovery and skin health.
- **TrueDark Glasses:** Blue light blocking eyewear for circadian rhythm alignment.

- **HumanCharger:** Bright light device targeting the brain via ear canal exposure.

12.2.5 Nutrition and Supplement Platforms

- **Bulletproof (Dave Asprey):** Well-known vendor for supplements, coffee, and biohacking products.
- **Thorne:** High-quality supplements with advanced lab testing.
- **Onnit:** Vendor offering nootropics, adaptogens, and fitness supplements.
- **Qualia (Neurohacker Collective):** Advanced nootropic stacks for cognitive enhancement.

12.2.6 Recovery and Stress-Management Tools

- **Apollo Neuro:** Wearable delivering gentle vibrations to improve stress resilience.
- **NuCalm:** Neuroacoustic software and wearable kit for deep relaxation.
- **Sensate:** Vagal nerve stimulation device for calming the nervous system.
- **Hyperice:** Recovery tools including percussion massage and compression devices.

12.2.7 Environmental and Body Optimization

- **ChiliSleep / Sleepme:** Cooling mattress pads to optimize sleep temperature.
- **Eight Sleep Pod:** Smart mattress with temperature regulation and sleep tracking.
- **AirDoctor:** Advanced air purifier for improved indoor air quality.
- **Hydrogen Water Machines:** Devices like Lourdes Hydrofix for antioxidant-rich water.

12.2.8 DIYbio and Community Labs

- **The ODIN:** Vendor for DIY biohacking kits and CRISPR learning tools.
- **BioCurious (California):** Community lab for DIY biology projects.
- **Genspace (New York):** Public biotech lab for citizen scientists.
- **Counter Culture Labs (Oakland):** Community-driven biotech and DIYbio hub.

12.2.9 Data Tracking and Analytics Platforms

- **InsideTracker:** Personalized blood biomarker analysis and nutrition insights.

- **Levels:** Continuous glucose monitoring (CGM) platform for metabolic health.
- **ZOE:** Microbiome and blood response testing for personalized nutrition.
- **Heads Up Health:** Platform for integrating and analyzing personal health data.

12.3 RECOMMENDED PROTOCOLS AND TRACKING TEMPLATES

- **Protocols for Nutrition and Metabolism:** Dietary interventions are among the most common forms of biohacking, and a structured protocol ensures meaningful results. A recommended approach begins with establishing baseline biomarkers such as blood glucose, lipid profiles, and body composition. Protocols should specify duration, type of diet (e.g., ketogenic, intermittent fasting, or plant-based), and timing of meals. Tracking templates should include columns for meal composition, caloric intake, timing, energy levels, digestive responses, and subjective mood. By recording this data daily, biohackers can correlate dietary changes with changes in metabolism, cognitive performance, and overall well-being. Protocols for supplementation follow similar structures, with detailed notes on dosage, timing, and observed effects.
- **Sleep Optimization Protocols:** Sleep is a cornerstone of health and performance, making it a key focus of biohacking protocols. Recommended protocols emphasize consistent sleep schedules, controlled light exposure, temperature optimization, and pre-sleep routines. Tracking templates for sleep should capture bedtime, wake-up time, total sleep duration, sleep quality ratings, use of aids such as blue-light blocking glasses, and environmental factors such as noise or temperature. Wearable devices or sleep trackers can supplement subjective ratings with objective measures like heart rate variability, REM cycles, and movement patterns. Analyzing this data over weeks allows biohackers to identify patterns and implement targeted interventions to improve restorative sleep.
- **Cognitive and Neuroenhancement Protocols:** Biohackers often experiment with cognitive enhancement strategies such as nootropics, meditation, neurofeedback, or brain stimulation devices. Protocols for cognitive biohacking start by establishing baseline cognitive performance through tests for memory, attention, or reaction time. Tracking templates record the intervention type, dosage (if applicable), duration, time of day, and subjective assessments of focus, mental

clarity, and mood. Objective metrics can include task performance, EEG readings, or digital cognitive assessments. Structured tracking allows biohackers to evaluate efficacy and detect unintended side effects, ensuring safer experimentation and clearer insights into which strategies are most effective.

- **Physical Performance and Recovery Protocols:** Enhancing physical performance and recovery is another area of biohacking that benefits from detailed protocols. Recommended approaches include baseline assessment of strength, endurance, flexibility, and cardiovascular health. Protocols should outline specific exercises, sets, repetitions, recovery periods, and any additional interventions such as cold exposure, sauna sessions, or compression therapy. Tracking templates should capture workout type, duration, intensity, perceived exertion, heart rate, and recovery metrics. Recovery-focused protocols also include sleep, hydration, nutrition, and biomarker tracking. Systematic documentation allows biohackers to understand how specific interventions influence both short-term performance and long-term adaptation.
- **Environmental and Sensory Optimization Protocols:** Environmental biohacking targets light exposure, temperature, air quality, and other sensory inputs that affect physiology and mood. Recommended protocols include consistent schedules for bright light exposure in the morning, limiting blue light in the evening, and monitoring indoor air quality or noise levels. Tracking templates should note exposure duration, environmental conditions, subjective energy or stress levels, and objective data from wearables when available. Grounding exercises or nature immersion protocols can be documented by logging time spent outdoors, activities performed, and corresponding changes in mood, stress, or focus. Structured observation helps biohackers understand the impact of subtle environmental changes on performance and well-being.
- **Genetic and Molecular Experimentation Protocols:** For advanced biohackers exploring genetic or molecular interventions, protocols emphasize strict safety, controlled experiments, and ethical considerations. Recommended steps include baseline genetic or microbiome assessments, defined intervention timelines, and careful documentation of reagents, dosages, and experimental conditions. Tracking templates should capture all variables, experimental observations, and any deviations from the protocol. Even minor details,

such as temperature fluctuations or procedural timings, can influence results. Accurate documentation ensures reproducibility, reduces risk, and provides a reliable record for evaluating long-term outcomes or sharing results with the scientific community.

- **Self-Tracking Templates:** Across all domains of biohacking, self-tracking templates provide a standardized way to capture data consistently. Effective templates include fields for date, intervention, dosage or duration, subjective ratings, objective metrics, and notes for unexpected outcomes. Digital tools such as spreadsheets, apps, or wearables can automate portions of data collection and generate graphs for trend analysis. Templates should be flexible to accommodate new experiments, additional biomarkers, or emerging technologies. Consistent use of tracking templates empowers biohackers to identify patterns, validate hypotheses, and make data-driven decisions in their self-experimentation.
- **Iteration and Analysis Protocols:** Biohacking relies on continuous iteration. Once initial data is collected, protocols should include steps for analyzing results, identifying correlations, and adjusting interventions. Tracking templates can support this process by summarizing trends, highlighting anomalies, and comparing multiple interventions. Biohackers are encouraged to review data weekly or monthly, refine experimental conditions, and document the rationale for changes. This iterative approach ensures that experiments evolve based on evidence rather than guesswork, improving both safety and effectiveness.
- **Ethical and Safety Considerations:** Protocols must incorporate ethical guidelines and safety measures. Clear documentation of experimental goals, potential risks, and safety precautions is essential, particularly for interventions involving supplements, environmental stressors, or molecular techniques. Tracking templates should include observations of adverse effects or unexpected outcomes. Recording these events responsibly protects the individual and the broader community, fostering a culture of responsible experimentation that balances innovation with well-being.

12.4 FURTHER READING: BOOKS, BLOGS, AND PODCASTS

12.4.1 Recommended Books

Books offer structured and in-depth exploration of biohacking concepts. One of the foundational texts is "Head Strong" by Dave Asprey, a seminal work on cognitive optimization and biohacking strategies for energy and mental performance.

Asprey, the founder of the Bulletproof brand, combines personal anecdotes with scientific insights to explain the impact of nutrition, sleep, and supplements on brain function. Another critical read is "Boundless," also by Asprey, which expands on longevity and performance optimization through advanced dietary strategies, supplements, and lifestyle interventions.

"Super Human" by Dave Asprey provides a detailed roadmap for achieving peak mental and physical performance, integrating topics such as nootropics, wearable technology, and metabolic optimization. Outside of Asprey's work, "Biohack Your Brain" by Kristen Willeumier focuses specifically on cognitive enhancement and neuroscience, providing actionable strategies for improving memory, focus, and emotional regulation. "The Biohacker's Handbook," authored by Teemu Arina, Olli Sovijärvi, and Jaakko Halmetoja, is a comprehensive guide covering nutrition, exercise, sleep, and supplementation, making it a practical reference for both beginners and experienced practitioners.

12.4.2 Popular Blogs

Blogs are valuable for staying current with emerging trends, experiments, and tools in biohacking. Ben Greenfield Fitness is a widely respected blog covering fitness, nutrition, recovery, and advanced biohacking technologies. Greenfield's content often combines scientific research with experiential insight, making it both informative and practical. Bulletproof Blog, maintained by Dave Asprey, provides guidance on cognitive performance, longevity, and lifestyle interventions, including detailed protocols and product reviews.

Other noteworthy blogs include SelfHacked, which focuses on health optimization using a combination of genetics, lifestyle, and supplementation. SelfHacked emphasizes evidence-based strategies and often references peer-reviewed research to support its claims. Quantified Self is a blog and community that highlights personal experiments, data tracking, and biofeedback strategies, appealing to individuals who prioritize data-driven self-optimization. These blogs often feature step-by-step guides, case studies, and analysis of new research, making them essential reading for anyone serious about biohacking.

12.4.3 Informative Podcasts

Podcasts offer a convenient way to learn about biohacking while on the go, combining interviews with experts, experiential insights, and discussions on emerging technologies. The Ben Greenfield Fitness Podcast is one of the most popular, covering topics from longevity and metabolic optimization to cognitive enhancement and wearable technology. Episodes often feature leading scientists, biohackers, and health professionals, providing diverse perspectives on self-optimization.

The Life Stylist Podcast, hosted by Luke Storey, explores biohacking alongside wellness, psychology, and spiritual growth. Storey interviews guests who share cutting-edge techniques for improving sleep, energy, and performance. The Biohacker Podcast, hosted by Anthony DiClementi, focuses specifically on trends, tools, and case studies within the biohacking community. This podcast often highlights emerging technologies, including brain-computer interfaces, nootropics, and genetic experimentation, giving listeners practical insights into experimental strategies.

Other influential podcasts include The Quantified Body and The Human Upgrade Podcast, which explore topics such as epigenetics, longevity, performance tracking, and advanced biohacking tools. These podcasts serve both as educational platforms and as inspiration for individuals experimenting with self-optimization.

12.4.4 Integration of Resources

Books, blogs, and podcasts complement each other in building a comprehensive understanding of biohacking. Books provide structured knowledge and long-form analysis, blogs offer practical step-by-step guidance and updates on emerging tools, and podcasts deliver dynamic discussions and expert interviews. Together, these resources equip biohackers with the knowledge to experiment safely, track results effectively, and stay informed about advancements in human performance and longevity. They also foster a sense of community, as biohackers can engage with authors, bloggers, and podcast hosts through social media, events, and collaborative experiments.

12.4.5 Emerging and Niche Resources

Beyond mainstream sources, several niche resources offer deep dives into specialized areas of biohacking. For example, InsideTracker Blog focuses on blood biomarker analysis and personalized health insights. ZOE Health provides research-driven insights into gut health and personalized nutrition, relevant for biohackers exploring microbiome optimization. Podcasts such as The Peter Attia Drive delve into longevity, metabolic health, and evidence-based performance strategies. These resources are particularly valuable for biohackers who prioritize data-driven approaches and scientifically rigorous experimentation.

12.5 LIST OF ONLINE BIOHACKING COMMUNITIES

- **Reddit Communities:** Reddit hosts several active biohacking subreddits where users discuss experiments, share research, and provide advice. **r/ Biohackers** is a general discussion community for DIY biology, cognitive enhancement, longevity, and wearable technology. **r/ Nootropics** focuses on cognitive enhancement substances, tracking personal experiences, and reviewing supplements. **r/QuantifiedSelf** is centered around self-tracking, data analytics, and metrics for personal optimization. Reddit allows anonymity and broad participation, making it ideal for sharing personal experiments and crowd-sourced insights.

- **Facebook Groups:** Facebook contains numerous private and public groups that foster discussion on biohacking topics. Groups like **Biohackers Global Community** and **DIYbio Enthusiasts** connect individuals across continents, facilitating peer learning, local meetups, and shared resources. These groups often post articles, tutorials, and event announcements, providing both practical guidance and opportunities for networking within the biohacking ecosystem.
- **Discord Servers:** Discord has become a popular hub for real-time biohacking discussions. Servers such as **The Biohacker Collective** host channels for different topics, including nutrition, genetics, neurofeedback, wearable technology, and longevity strategies. Members can ask questions, share data, and discuss recent scientific studies. Voice channels allow interactive workshops and live experiment walkthroughs, fostering a collaborative and engaging environment.
- **DIY Biology Labs and Networks:** Several community labs operate online platforms for members to connect, share protocols, and access virtual resources. **BioCurious**, based in California, maintains forums and mailing lists for DIYbio enthusiasts. **Counter Culture Labs** in Oakland offers online collaboration spaces and virtual workshops alongside their physical lab. **Genspace** in New York provides educational webinars and discussion boards that allow remote participation in biohacking experiments. These networks combine physical lab access with online communities to democratize biological experimentation.
- **Quantified Self Community:** The **Quantified Self** organization offers an online community platform where members share self-tracking experiments, methodologies, and results. Forums and virtual meetups focus on wearable technology, data visualization, and experimental protocols for personal optimization. The platform emphasizes evidence-based experimentation, encouraging participants to share both successes and failures for collective learning.
- **Slack and Professional Networks:** Some biohackers use professional-oriented communication tools like Slack to discuss high-level topics such as synthetic biology, CRISPR experiments, and corporate biohacking initiatives. Slack communities such as **DIYbio Network** and **Longevity**

Biohacker Hub offer moderated channels, document sharing, and expert Q&A sessions. These communities cater to individuals seeking more structured discussions, mentorship, and professional collaboration opportunities.

- **Podcasts and Interactive Forums:** Several biohacking podcasts have companion online communities that extend discussion beyond the episodes. For example, **The Ben Greenfield Fitness Podcast** and **The Human Upgrade Podcast** maintain Facebook groups and private forums where listeners discuss protocols, share outcomes, and exchange practical advice. These communities integrate media content with interactive engagement, helping members implement strategies covered in the shows.
- **International Communities:** Biohacking is a global phenomenon, and many international online communities cater to local languages and regulations. For example, **HackBio (Europe)** offers discussion boards and virtual meetups in multiple European languages, while **Biohacking India Community** connects enthusiasts within India for online workshops, webinars, and peer-to-peer mentorship. These networks highlight the cultural diversity of biohacking while maintaining a focus on shared experimentation and innovation.
- **Safety and Ethical Considerations in Online Communities:** While online communities are invaluable, they also require vigilance regarding safety and ethics. Discussions often involve self-experimentation, supplements, or unregulated interventions. Members are encouraged to share disclaimers, cite sources, and emphasize evidence-based practices. Communities with moderation policies, rules for sharing experimental data, and expert oversight tend to foster safer and more responsible biohacking practices.

References

Abundance: The Future Is Better Than You Think, Author: Peter H. Diamandis, Steven Kotler, Publication: Free Press, Publish Year: 2012

Atomic Habits: An Easy & Proven Way to Build Good Habits & Break Bad Ones, Author: James Clear, Publication: Avery, Publish Year: 2018

Bad Blood: Secrets and Lies in a Silicon Valley Startup, Author: John Carreyrou, Publication: Knopf, Publish Year: 2018

Benjamin Franklin: An American Life, Author: Walter Isaacson, Publication: Simon & Schuster, Publish Year: 2003

Beyond Training: Mastering Endurance, Health & Life, Author: Ben Greenfield, Publication: Victory Belt Publishing, Publish Year: 2014

Biohacker's Handbook, The: A Step-by-Step Guide to Upgrade Your Body, Mind and Health, Author: Teemu Arina, Olli Sovijärvi, Jaakko Halmetoja, Publication: Biohacker Center, Publish Year: 2018

Bold: How to Go Big, Create Wealth and Impact the World, Author: Peter H. Diamandis, Steven Kotler, Publication: Simon & Schuster, Publish Year: 2015

Boundless: Upgrade Your Brain, Optimize Your Body & Defy Aging, Author: Ben Greenfield, Publication: Victory Belt Publishing, Publish Year: 2020

Brain Maker: The Power of Gut Microbes to Heal and Protect Your Brain–for Life, Author: David Perlmutter, Publication: Little, Brown Spark, Publish Year: 2015

Can't Hurt Me: Master Your Mind and Defy the Odds, Author: David Goggins, Publication: Lioncrest Publishing, Publish Year: 2018

Change Your Schedule, Change Your Life: How to Harness the Power of Clock Genes to Lose Weight, Optimize Your Workout, and Finally Get a Good Night's Sleep, Author: Dr. Suhas Kshirsagar, Michelle D. Seaton, Publication: HarperOne, Publish Year: 2018

Deep Work: Rules for Focused Success in a Distracted World, Author: Cal Newport, Publication: Grand Central Publishing, Publish Year: 2016

Digital Minimalism: Choosing a Focused Life in a Noisy World, Author: Cal Newport, Publication: Portfolio, Publish Year: 2019

Einstein: His Life and Universe, Author: Walter Isaacson, Publication: Simon & Schuster, Publish Year: 2007

Elon Musk: Tesla, SpaceX, and the Quest for a Fantastic Future, Author: Ashlee Vance, Publication: Ecco, Publish Year: 2015

Everything Is F*cked: A Book About Hope, Author: Mark Manson, Publication: Harper, Publish Year: 2019

Flow: The Psychology of Optimal Experience, Author: Mihaly Csikszentmihalyi, Publication: Harper & Row, Publish Year: 1990

Genius Foods: Become Smarter, Happier, and More Productive While Protecting Your Brain for Life, Author: Max Lugavere, Publication: Harper Wave, Publish Year: 2018

Go Wild: Free Your Body and Mind from the Afflictions of Civilization, Author: John J. Ratey, Richard Manning, Publication: Little, Brown Spark, Publish Year: 2014

Grain Brain: The Surprising Truth about Wheat, Carbs, and Sugar–Your Brain's Silent Killers, Author: David Perlmutter, Publication: Little, Brown Spark, Publish Year: 2013

Grit: The Power of Passion and Perseverance, Author: Angela Duckworth, Publication: Scribner, Publish Year: 2016

Head Strong: The Bulletproof Plan to Activate Untapped Brain Energy to Work Smarter and Think Faster-in Just Two Weeks, Author: Dave Asprey, Publication: Harper Wave, Publish Year: 2017

Homo Deus: A Brief History of Tomorrow, Author: Yuval Noah Harari, Publication: Harper, Publish Year: 2017

How to Create a Mind: The Secret of Human Thought Revealed, Author: Ray Kurzweil, Publication: Viking, Publish Year: 2012

How to Win Friends and Influence People, Author: Dale Carnegie, Publication: Simon & Schuster, Publish Year: 1936

In the Realm of Hungry Ghosts: Close Encounters with Addiction, Author: Gabor Maté, Publication: Knopf Canada, Publish Year: 2008

Influence: The Psychology of Persuasion, Author: Robert B. Cialdini, Publication: Harper Business, Publish Year: 1984

Leonardo da Vinci, Author: Walter Isaacson, Publication: Simon & Schuster, Publish Year: 2017

Lifespan: Why We Age—and Why We Don't Have To, Author: David A. Sinclair, Publication: Atria Books, Publish Year: 2019

Mastery, Author: Robert Greene, Publication: Viking, Publish Year: 2012

Mindset: The New Psychology of Success, Author: Carol S. Dweck, Publication: Random House, Publish Year: 2006

Nudge: Improving Decisions About Health, Wealth, and Happiness, Author: Richard H. Thaler, Cass R. Sunstein, Publication: Yale University Press, Publish Year: 2008

Predictably Irrational: The Hidden Forces That Shape Our Decisions, Author: Dan Ariely, Publication: HarperCollins, Publish Year: 2008

Pre-Suasion: A Revolutionary Way to Influence and Persuade, Author: Robert B. Cialdini, Publication: Simon & Schuster, Publish Year: 2016

Sapiens: A Brief History of Humankind, Author: Yuval Noah Harari, Publication: Harper, Publish Year: 2015

Shoe Dog: A Memoir by the Creator of Nike, Author: Phil Knight, Publication: Scribner, Publish Year: 2016

So Good They Can't Ignore You: Why Skills Trump Passion in the Quest for Work You Love, Author: Cal Newport, Publication: Grand Central Publishing, Publish Year: 2012

Spark: The Revolutionary New Science of Exercise and the Brain, Author: John J. Ratey, Publication: Little, Brown Spark, Publish Year: 2008

Stealing Fire: How Silicon Valley, the Navy SEALs, and Maverick Scientists Are Revolutionizing the Way We Live and Work, Author: Steven Kotler, Jamie Wheal, Publication: Dey Street Books, Publish Year: 2017

Steve Jobs, Author: Walter Isaacson, Publication: Simon & Schuster, Publish Year: 2011

Super Human: The Bulletproof Plan to Age Backward and Maybe Even Live Forever, Author: Dave Asprey, Publication: Harper Wave, Publish Year: 2019

The 10X Rule: The Only Difference Between Success and Failure, Author: Grant Cardone, Publication: Wiley, Publish Year: 2010

The 48 Laws of Power, Author: Robert Greene, Publication: Viking, Publish Year: 1998

The 4-Hour Body: An Uncommon Guide to Rapid Fat-Loss, Incredible Sex, and Becoming Superhuman, Author: Timothy Ferriss, Publication: Harmony, Publish Year: 2010

The 5 AM Club: Own Your Morning, Elevate Your Life, Author: Robin Sharma, Publication: HarperCollins, Publish Year: 2018

The 7 Habits of Highly Effective People: Powerful Lessons in Personal Change, Author: Stephen R. Covey, Publication: Free Press, Publish Year: 1989

The Art of Learning: An Inner Journey to Optimal Performance, Author: Josh Waitzkin, Publication: Free Press, Publish Year: 2007

The Biology of Belief: Unleashing the Power of Consciousness, Matter & Miracles, Author: Bruce H. Lipton, Publication: Hay House, Publish Year: 2005

The Body Keeps the Score: Brain, Mind, and Body in the Healing of Trauma, Author: Bessel van der Kolk, Publication: Viking, Publish Year: 2014

The Circadian Code: Lose Weight, Supercharge Your Energy, and Transform Your Health from Morning to Midnight, Author: Satchin Panda, Publication: Rodale Books, Publish Year: 2018

The Complete Guide to Biohacking, Author: Ron Kness, Publication: N/A, Publish Year: 2022

The Compound Effect, Author: Darren Hardy, Publication: Success Books, Publish Year: 2010

The Daniel Plan: 40 Days to a Healthier Life, Author: Rick Warren, Daniel Amen, Mark Hyman, Publication: Zondervan, Publish Year: 2013

The Emperor of All Maladies: A Biography of Cancer, Author: Siddhartha Mukherjee, Publication: Scribner, Publish Year: 2010

The End of Alzheimer's: The First Program to Prevent and Reverse Cognitive Decline, Author: Dale Bredesen, Publication: Avery, Publish Year: 2017

The Everything Store: Jeff Bezos and the Age of Amazon, Author: Brad Stone, Publication: Little, Brown and Company, Publish Year: 2013

The Future Is Faster Than You Think: How Converging Technologies Are Transforming Business, Industries, and Our Lives, Author: Peter H. Diamandis, Steven Kotler, Publication: Simon & Schuster, Publish Year: 2020

The Gene: An Intimate History, Author: Siddhartha Mukherjee, Publication: Scribner, Publish Year: 2016

The Genius Life: Heal Your Mind, Strengthen Your Body, and Become Extraordinary, Author: Max Lugavere, Publication: Harper Wave, Publish Year: 2020

The Happiness Hypothesis: Finding Modern Truth in Ancient Wisdom, Author: Jonathan Haidt, Publication: Basic Books, Publish Year: 2006

The Hormone Cure: Reclaim Balance, Sleep and Sex Drive; Lose Weight; Feel Focused, Vital, and Energized Naturally with the Gottfried Protocol, Author: Sara Gottfried, Publication: Scribner, Publish Year: 2013

The Hormone Reset Diet: Heal Your Metabolism to Lose Up to 15 Pounds in 21 Days, Author: Sara Gottfried, Publication: HarperOne, Publish Year: 2015

The Innovators: How a Group of Hackers, Geniuses, and Geeks Created the Digital Revolution, Author: Walter Isaacson, Publication: Simon & Schuster, Publish Year: 2014

The Joy of Movement: How Exercise Helps Us Find Happiness, Hope, Connection, and Courage, Author: Kelly McGonigal, Publication: Avery, Publish Year: 2019

The Laws of Human Nature, Author: Robert Greene, Publication: Viking, Publish Year: 2018

The Lean Startup: How Today's Entrepreneurs Use Continuous Innovation to Create Radically Successful Businesses, Author: Eric Ries, Publication: Crown Business, Publish Year: 2011

The Longevity Paradox: How to Die Young at a Ripe Old Age, Author: Steven R. Gundry, Publication: Harper Wave, Publish Year: 2019

The Magic of Thinking Big, Author: David J. Schwartz, Publication: Prentice Hall Press, Publish Year: 1959

The Metabolic Approach to Cancer: Integrating Deep Nutrition, the Ketogenic Diet, and Nontoxic Bio-Individualized Therapies, Author: Dr. Nasha Winters, Jess Higgins Kelley, Publication: Chelsea Green Publishing, Publish Year: 2017

The One Thing: The Surprisingly Simple Truth Behind Extraordinary Results, Author: Gary Keller, Jay Papasan, Publication: Bard Press, Publish Year: 2012

The Oxygen Advantage: The Simple, Scientifically Proven Breathing Techniques for a Healthier, Slimmer, Faster, and Fitter You, Author: Patrick McKeown, Publication: William Morrow Paperbacks, Publish Year: 2015

The Paleo Solution: The Original Human Diet, Author: Robb Wolf, Publication: Victory Belt Publishing, Publish Year: 2010

The Paradox of Choice: Why More Is Less, Author: Barry Schwartz, Publication: Ecco, Publish Year: 2004

The Plant Paradox: The Hidden Dangers in "Healthy" Foods That Cause Disease and Weight Gain, Author: Steven R. Gundry, Publication: Harper Wave, Publish Year: 2017

The Power Broker: Robert Moses and the Fall of New York, Author: Robert A. Caro, Publication: Knopf, Publish Year: 1974

The Power of Now: A Guide to Spiritual Enlightenment, Author: Eckhart Tolle, Publication: New World Library, Publish Year: 1997

The Relaxation Response, Author: Herbert Benson, Publication: HarperTorch, Publish Year: 1975

The Righteous Mind: Why Good People Are Divided by Politics and Religion, Author: Jonathan Haidt, Publication: Pantheon, Publish Year: 2012

The Rise of Superman: Decoding the Science of Ultimate Human Performance, Author: Steven Kotler, Publication: New Harvest, Publish Year: 2014

The Science of Getting Rich, Author: Wallace D. Wattles, Publication: TarcherPerigee, Publish Year: 1910

The Selfish Gene, Author: Richard Dawkins, Publication: Oxford University Press, Publish Year: 1976

The Singularity Is Near: When Humans Transcend Biology, Author: Ray Kurzweil, Publication: Viking, Publish Year: 2005

The Sleep Revolution: Transforming Your Life, One Night at a Time, Author: Arianna Huffington, Publication: Harmony, Publish Year: 2016

The Subtle Art of Not Giving a F*ck: A Counterintuitive Approach to Living a Good Life, Author: Mark Manson, Publication: HarperOne, Publish Year: 2016

The Third Wave: An Entrepreneur's Vision of the Future, Author: Steve Case, Publication: Simon & Schuster, Publish Year: 2016

The UltraMind Solution: Fix Your Broken Brain by Healing Your Body First, Author: Mark Hyman, Publication: Scribner, Publish Year: 2008

The Wedge: Evolution, Consciousness, Stress, and the Key to Human Resilience, Author: Scott Carney, Publication: Rodale Books, Publish Year: 2020

The Willpower Instinct: How Self-Control Works, Why It Matters, and What You Can Do to Get More of It, Author: Kelly McGonigal, Publication: Avery, Publish Year: 2011

The Wright Brothers, Author: David McCullough, Publication: Simon & Schuster, Publish Year: 2015

The Years of Lyndon Johnson (4 Volumes), Author: Robert A. Caro, Publication: Knopf, Publish Year: 1982-2012

Think and Grow Rich, Author: Napoleon Hill, Publication: The Ralston Society, Publish Year: 1937

Thinking, Fast and Slow, Author: Daniel Kahneman, Publication: Farrar, Straus and Giroux, Publish Year: 2011

Tools of Titans: The Tactics, Routines, and Habits of Billionaires, Icons, and World-Class Performers, Author: Timothy Ferriss, Publication: Houghton Mifflin Harcourt, Publish Year: 2016

What Doesn't Kill Us: How Freezing Water, Extreme Altitude, and Environmental Conditioning Will Renew Our Lost Evolutionary Strength, Author: Scott Carney, Publication: Rodale Books, Publish Year: 2017

When the Body Says No: The Cost of Hidden Stress, Author: Gabor Maté, Publication: Knopf Canada, Publish Year: 2003

Why We Sleep: Unlocking the Power of Sleep and Dreams, Author: Matthew Walker, Publication: Scribner, Publish Year: 2017

Wired to Eat: Turn Off Cravings, Rewire Your Appetite for Weight Loss, and Determine the Foods That Work for You, Author: Robb Wolf, Publication: Harmony, Publish Year: 2017

Zero to One: Notes on Startups, or How to Build the Future, Author: Peter Thiel, Blake Masters, Publication: Crown Business, Publish Year: 2014

Index

L

M

N

O

P

Q

www.ingramcontent.com/pod-product-compliance
Lightning Source LLC
LaVergne TN
LVHW081937200926
848484LV00009B/2455
* 9 7 8 1 7 8 7 1 5 6 6 0 9 *